NASA STI Program . . . in Profile

Since its founding, NASA has been dedicated to the advancement of aeronautics and space science. The NASA Scientific and Technical Information (STI) program plays a key part in helping NASA maintain this important role.

The NASA STI Program operates under the auspices of the Agency Chief Information Officer. It collects, organizes, provides for archiving, and disseminates NASA's STI. The NASA STI program provides access to the NASA Aeronautics and Space Database and its public interface, the NASA Technical Reports Server, thus providing one of the largest collections of aeronautical and space science STI in the world. Results are published in both non-NASA channels and by NASA in the NASA STI Report Series, which includes the following report types:

- TECHNICAL PUBLICATION. Reports of completed research or a major significant phase of research that present the results of NASA programs and include extensive data or theoretical analysis. Includes compilations of significant scientific and technical data and information deemed to be of continuing reference value. NASA counterpart of peer-reviewed formal professional papers but has less stringent limitations on manuscript length and extent of graphic presentations.

- TECHNICAL MEMORANDUM. Scientific and technical findings that are preliminary or of specialized interest, e.g., quick release reports, working papers, and bibliographies that contain minimal annotation. Does not contain extensive analysis.

- CONTRACTOR REPORT. Scientific and technical findings by NASA-sponsored contractors and grantees.

- CONFERENCE PUBLICATION. Collected papers from scientific and technical conferences, symposia, seminars, or other meetings sponsored or cosponsored by NASA.

- SPECIAL PUBLICATION. Scientific, technical, or historical information from NASA programs, projects, and missions, often concerned with subjects having substantial public interest.

- TECHNICAL TRANSLATION. English-language translations of foreign scientific and technical material pertinent to NASA's mission.

Specialized services also include creating custom thesauri, building customized databases, organizing and publishing research results.

For more information about the NASA STI program, see the following:

- Access the NASA STI program home page at *http://www.sti.nasa.gov*

- E-mail your question via the Internet to *help@sti.nasa.gov*

- Fax your question to the NASA STI Help Desk at 301–621–0134

- Telephone the NASA STI Help Desk at 301–621–0390

- Write to:
  NASA Center for AeroSpace Information (CASI)
  7115 Standard Drive
  Hanover, MD 21076–1320

Additional Technologies and Investigations for Provision of Future Aeronautical Communications

Tricia Gilbert, Jenny Jin, Jason Berger, and Steve Henriksen
ITT Corporation, Herndon, Virginia

Prepared for the
Aeronautics Communications Panel (ACP), Working Group T–1
sponsored by the International Civil Aviation Organization (ICAO)
Montreal, Canada, September 19–21, 2007

Prepared under Contract NNC05CA85C

National Aeronautics and
Space Administration

Glenn Research Center
Cleveland, Ohio 44135

February 2008

Level of Review: This material has been technically reviewed by NASA expert reviewer.

Available from

NASA Center for Aerospace Information
7115 Standard Drive
Hanover, MD 21076–1320

National Technical Information Service
5285 Port Royal Road
Springfield, VA 22161

Preface

The following NASA Contractor Report documents the indepth studies on select technologies that could support long-term aeronautical mobile communications operating concepts. This work was performed during the third and final phase of NASA's Technology Assessment for the FAA/ EUROCONTROL Future Communications Study (FCS) under a multiyear NASA contract. It includes the associated findings of ITT Corporation and NASA Glenn Research Center to the Federal Aviation Administration (FAA) as of the end of May 2007. The activities documented in this report focus on three final technology candidates identified by the United States, and were completed before sufficient information about two additional technology candidates proposed by EUROCONTROL was made available. A separate report to be published by NASA (NASA/CR—2008-215144), entitled "Final Report on Technology Investigations for Provision of Future Aeronautical Communications" will include an assessment of all five final candidate technologies considered by the U.S. agencies (FAA and NASA) and EUROCONTROL. It will also provide an overview of the entire technology assessment process, including final recommendations. All three phases of this work were performed in compliance with the Terms of Reference for the Action Plan number 17 (AP–17) cooperative research agreement among EUROCONTROL, FAA, and NASA along with the general guidance of the FAA and EUROCONTROL available throughout this study.

Executive Summary

E.S.1 Background and Introduction

The Future Communication Study (FCS) is a cooperative research and development program of the Federal Aviation Administration (FAA), National Aeronautics and Space Administration (NASA), and EUROCONTROL. This study has several technical themes supporting the definition of a future globally interoperable communications system to support air traffic management (ATM) operations in the timeframe of 2020 and beyond. One of these themes calls for "investigation of potential communications technologies operating inside the very high frequency (VHF) band and outside the VHF band to support the long-term mobile communication operation concept considering terrestrial and satellite base infrastructure." The Future Communication Study has been organized and carried out in three phases: Technology Prescreening (Phase I), Technology Screening and Indepth Studies (Phase II), and Additional Technologies and Investigations for Provision of Future Aeronautical Communications (Phase III).

As decision-making in the aeronautical environment can be complex, a structured methodology that accommodates stakeholder inputs has been defined and applied in the FCS. This approach is shown in figure 1. The first set of activities (steps 1A and 1B) included derivation of evaluation criteria and metrics. Step 2 is to identify most promising technology candidates. The remaining steps (steps 3 through 6) contribute to detailed assessment of the most promising candidate technologies. Indepth analysis of the considered technologies supported these steps. The process concludes with consideration of the relative importance of criteria to identify the best performing technology.

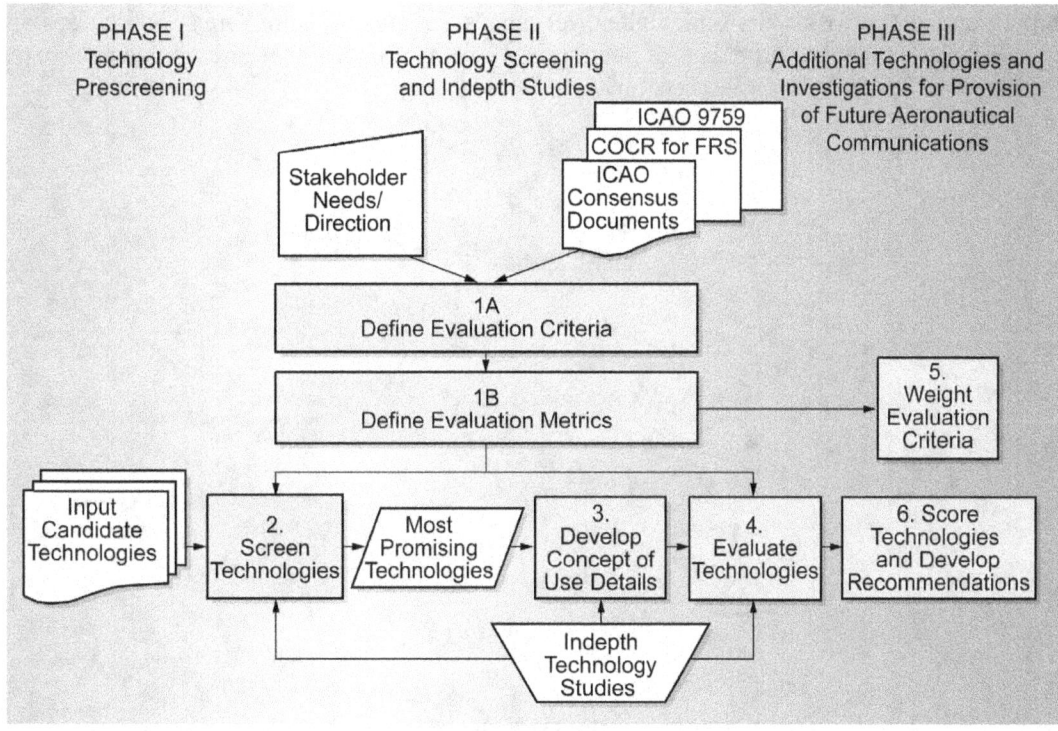

Figure 1.—FCS technology investigation methodology.

This report documents the indepth technology analysis activities and results that were conducted during the third phase of the technology investigation (Phase III). The FCS final report will document final technology evaluation results and recommendations.

E.S.2 Phase III Indepth Studies

Three indepth studies were conducted as part of the FCS Phase III study efforts to support the technology evaluation process and to better understand the applicability of the most promising technologies to the future aeronautical communication environment; they are the L-band interference testing, the wideband code division multiple access (WCDMA) functional assessment, and the P34/TIA(Telecommunications Industry Association)–902 intellectual property assessment.

E.S.2.1 The L-band Interference Testing

The focus of the L-band interference testing was to gain an understanding of the L-band interference environment and to measure interference for three candidate technologies against the distance measure equipment (DME), a navigation system currently operating in the aeronautical L-band spectrum. Before any new communication systems can be allowed to share spectrum with DME, a compatibility analysis must assess potential degradation of DME system performance. A compatibility analysis of candidate technologies with DME in L-band involves many potential interference scenarios, including cosite (onboard the aircraft), air-to-air, air-to-ground/ground-to-air, and ground-to-ground.

For the interference measurements of this study, the cosite interference scenario was emulated. The cosite interference scenario serves as a guide for specifying the bounds of interference power used when taking measurements. The signal levels used for interference measurements should encompass the power ranges that will be seen in practice, and these levels are greatest in the cosite scenario. This is not to say that ground stations do not also transmit high power signals, but ground stations are generally separated by large distances to mitigate the effects of interference. Antennae onboard the aircraft do not have this liberty. Figure 2 illustrates the cosite scenario.

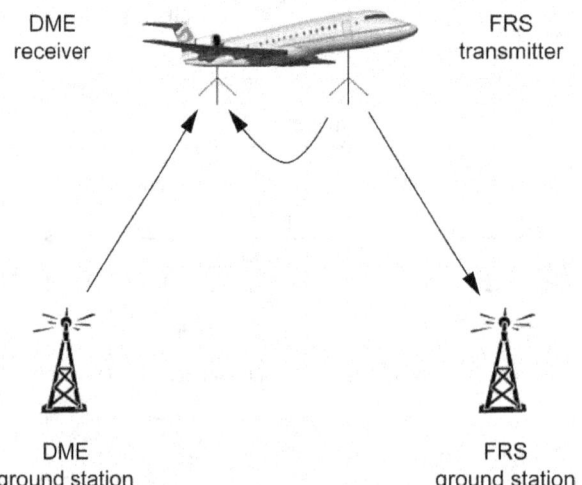

DME
receiver

FRS
transmitter

DME
ground station

FRS
ground station

Figure 2.—Cosite interference scenario.

Results of the study improved understanding of the interference environment associated with DME. While limitations in the test equipment do not provide enough information to address specific channelization techniques for future radio system (FRS) candidates, the interference measurements do suggest that the addition of an FRS in the DME band may be feasible. Three major findings of the interference measurements study are as follows:

- The power levels expected from continuously transmitting FRS equipment onboard the aircraft may be sufficiently high as to cause desensitization in the DME interrogator. This phenomenon was evident for all of the FRS candidates, even at large frequency separations for the DME that was tested. This finding is not favorable for FRS candidate technologies whose concept of use assumes continuous transmissions (e.g., WCDMA).

- The data also indicate that the DME interrogator is more tolerant to gated transmissions (i.e., there is potential for implementation of a technology with a gated waveform; but off-set channels may still be required (to be investigated)). A majority of the measurements used 100 percent duty cycles, which results in a conservative analysis. Lesser duty cycles may be expected in practice. It is expected that low-to-moderate duty cycles will interfere less with DME compared to FRSs with high duty cycles. This finding may be favorable for FRS candidate technologies whose concept of use assumes noncontinuous transmissions (i.e., L-band digital link (LDL) and P34/TIA–902 (partial)).

- Further analysis should be conducted to characterize the relationship between FRS duty-cycle and interference susceptibility (the duty-cycle investigation should include more variables than just overall duty cycle; there may be some combination of specific timescales of on/off pulses and overall duty cycle that results in a seemingly "invisible" waveform from the DME interrogator's perspective.) In the context of this investigation, identification of collocation constraints can also be investigated. It is also recommended that different models of DME interrogators be tested to provide a range of performance.

The interference measurements, in conjunction with the concept of use defined for each candidate FRS technology, will be used to support the detailed technology evaluations to be documented in the FCS final report.

E.S.2.2 The WCDMA Functional Assessment

A second indepth analysis was the WCDMA Functional Assessment. The study's objective was to define how the technology could support Communications Operating Concept and Requirements (COCR) services. Specifically, the analysis was used to determine the necessary elements of the architecture and protocol stack to provision COCR services, which further supported the assessment of cost, certification, and standardization of WCDMA in aeronautical applications. A representative result of this analysis, mapping of one part of a COCR service, data link initiation capability (DLIC), to WCDMA, is illustrated in figure 3.

Applying the WCDMA standards as defined, this study found that a full complement of WCDMA functional elements is required to provision COCR services. Not only the air interface and elements of the radio network controller are needed, but also elements of the core network such as home location register (HLR), serving general packet radio service (GPRS), support node (SGSN), and gateway GPRS support node (GGSN).

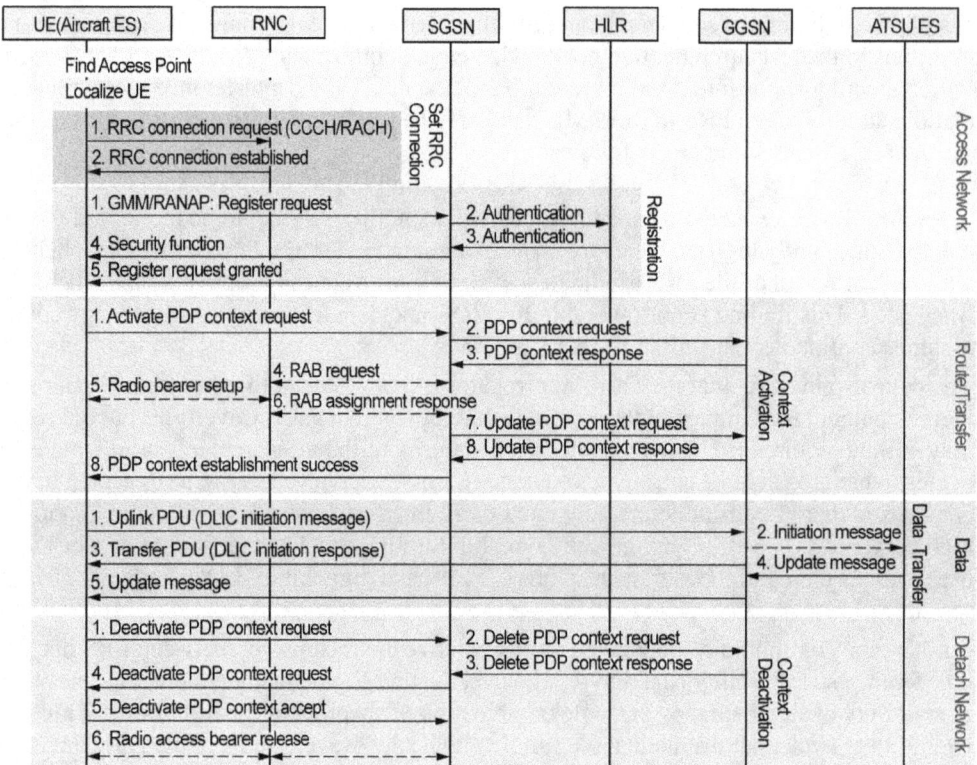

Figure 3.—Map DLIC Initation+Update functions to WCDMA service.

E.S.2.3 The P34/TIA–902 Intellectual Property Assessment

The final indepth analysis conducted during Phase III of the FCS technology investigation was the P34/TIA–902 intellectual property assessment. This study evaluated the potential impact of the P34/TIA–902 standard intellectual property in the context of an FRS implementation. Thorough review of the patents and consultation with a patent counsel helped develop the following conclusions:

- The concept of use defined for P34/TIA–902 makes some patents not applicable (e.g., isotropic orthogonal transform algorithm (IOTA) physical layer not used in the FCS application and associated patents do not apply); also, recommended tailoring of physical layer standard for the FCS application results in bypassing of most physical layer patents.

- Only one patent is assessed as desirable to implement. It is a methodology proposed for power amplifier linearization, modification of which would influence the definition of MAC framing structure.

- Most, if not all, patents will expire before timeframe of FCS.

- These patents are not applicable to companies outside the United States.

- Intellectual property associated with P34/TIA–902 standard is deemed to have little or no impact on the FRS if it is an implementation based on this standard.

Contents

1.0 Background and Introduction

1.1 Global Aeronautical Communication Objectives

The origin of current aeronautical communication objectives can be traced to results of the International Civil Aviation Organization (ICAO) Aeronautical Mobile Communication Panel (AMCP) from the year 2000 and the Eleventh Air Navigation Conference (ANConf/11), held in Montréal, Canada, from September 22 through October 3, 2003. One result of the seventh meeting of the AMCP (AMCP/7) in March 2000 was the establishment of a task (Task communication, surveillance, navigation (CNS)–9102) to carry out the fact-finding and conduct the necessary studies for the development of data links for air traffic services and aeronautical operational control. In October 2000, the AMCP working group C (WG–C), addressing future air/ground communications, held their first meeting, which included the establishment of Action WGC/1–9 to develop a report to recommend a scenario in which a common global interoperable communication infrastructure could be ensured for the future. Finally, one of the highlights of the formal ICAO Air Navigation Conference was the official report of the Technical and Operational Matters in Air Traffic Control Committee (Committee B). This report noted the current state of aviation communications and made several recommendations to advance this state. The observations included

- The aeronautical mobile communication infrastructure has to evolve in order to accommodate new functions.
- This evolution would likely require the definition and implementation of new terrestrial and/or satellite systems that operate outside the very high frequency (VHF) band.
- A variety of (somewhat divergent) views had been presented with regard to the future evolution of aeronautical mobile communications.
- The universally recognized benefits of harmonization and global interoperability of air ground communications should not be forgotten when pursuing optimized local solutions.
- The successful gradual introduction of data communications should be continued to complement and replace voice for routine communications.

Based on these observations, several conference recommendations were made. These included

- Recommendation 7/4—Investigation of future technology alternatives for air-ground communications. That ICAO
  - investigate new terrestrial and satellite-based technologies, on the basis of their potential for ICAO standardization for aeronautical mobile communications use, taking into account the safety-critical standards of aviation and the associated cost issues.
- Recommendation 7/5—Standardization of aeronautical communication systems. That, for new aeronautical communication systems, ICAO
  - continue to monitor emerging communication systems technologies but undertake standardization work only when the systems meet all of the following conditions:
    - o can meet current and emerging ICAO ATM requirements
    - o are technically proven and offer proven operational benefits
    - o are consistent with the requirements for safety
    - o are cost beneficial
    - o can be implemented without prejudice to global harmonization of the CNS/ATM systems
    - o are consistent with the Global Air Navigation Plan for CNS/ATM Systems (Doc. 9750)

At ANConf/11, there was a strong request particularly from the airlines (International Air Transport Association) for international cooperation in order to achieve the stated objectives and goals in a harmonized and globally interoperable manner. In part to address the ICAO actions and recommendations above and in part to address frequency congestion and spectrum depletion in core Europe and dense United States airspace, the FAA and EUROCONTROL have embarked on a cooperative research and development program. The terms of this program are outlined in the Terms of Reference document for the program, which has been entitled "Future Communications Study (FCS)". By agreement, joint FAA and EUROCONTROL research and development activities require terms of reference, which are referred to as "action plans" and are numbered sequentially. The Terms of Reference for the Future Communications Study are detailed in Action Plan 17 (AP–17), and NASA, the FAA, and EUROCONTROL all have defined roles in the research and development activities.

1.2   Future Communication Study Technology Investigations

The terms of reference for the FCS organized the work program into six technical and three business themes supporting the definition of a future globally interoperable communications system for air traffic management (ATM) operations in the timeframe of 2020 and beyond. Three of the technical themes address key activities relating to the identification of the most suitable technology candidates for the future communication infrastructure. These include (1) identification of requirements and operating concepts, (2) technology alternatives assessment, and (3) development of a future communications roadmap.

The first theme has been addressed through the development of the Communications Operating Concept and Requirements (COCR) for the future radio system (FRS), a document which describes the future operating concepts and environment associated with safety and regularity of flight including air traffic services (ATS) and safety-related aeronautical operational control (AOC) communications. The document also describes the operational and communication requirements associated with the radio components of a communication system collectively referred to as FRS. The second theme applies the material captured in the COCR to perform a technology assessment. Specifically, the FCS terms of reference calls for "investigation of potential communications technologies operating inside the VHF band and outside the VHF band to support the long-term mobile communication operation concept considering terrestrial and satellite base infrastructure." Finally, based on current and planned operational aeronautical communication systems both within core Europe and the United States, and considering results of the technology assessment, the final technical theme includes a definition of a communications roadmap that supports "planning for and achieving smooth transition" to recommended technologies.

NASA was to provide the leading role in this effort. Specific investigations have been performed as a sequence of three study phases, Technology Prescreening (Phase I), Technology Screening and Indepth Studies (Phase II), and Additional Technologies and Investigations for Provision of Future Aeronautical Communications (Phase III). Interim reports associated with results of the first two study phases are available as follows:

- Phase I (completed December 2004): "Technology Assessment for the Future Aeronautical Communications System," NASA/CR—2005-213587, available at http://gltrs.grc.nasa.gov/reports/2005/CR-2005-213587.pdf
- Phase II (completed July 2006): "Identification of Technologies for Provision of Future Aeronautical Communications," NASA/CR—2006-214451 (ref. 3), available at http://gltrs.grc.nasa.gov/reports/2006/CR-2006-214451.pdf

1.3   Stakeholder Inputs

During the course of the FCS, interim findings were briefed to FAA and EUROCONTROL senior management, ICAO, industry, and to the United States Air Traffic Management Advisory Committee

(ATMAC). Significant feedback was received on some of the interim study results. Some raised concern on moving to a new communication band due to perceived cost ramifications of additional ground infrastructure and either additional hull penetrations or costly equipment integration on aircraft. ATMAC defined a set of recommendations that related to future aeronautical communication capability that included (ref. 1)

- Sustain voice in VHF spectrum as long as possible, maintaining analog 25-kHz double side band-amplitude modulation (DSB–AM) until such time as spectrum pressures require reducing channel spacing to 8.33 kHz.
- Pursue new technological solutions as a last resort.
- Data link is important—commit to a technology and implement by 2015.
- Keep AOC and ATS separate.

The FAA indicated intent to comply with the ATMAC recommendations, but also a desire to prepare for the future. Should the capacity of the aeronautical VHF spectrum ever prove insufficient to provide the total data link capacity required, the FAA would support a new system to be ready and available to ensure that the aviation communications needs are accommodated. This is completely in line with the ICAO ANConf/11 observation that "This evolution would likely require the definition and implementation of new terrestrial and/or satellite systems that operate outside the VHF band." This same theme was reflected by EUROCONTROL, which indicated that European focus is on consideration of an L-band system. EUROCONTROL also explored the potential of satellite systems, however initial analysis concluded that availability may preclude their use as a primary system in continental airspace. After feedback on Phase I results, subsequent FCS technology investigation focus has been made to support the understanding of issues associated with hosting a communications system in either L-band or C-band and with the potential use of satellites for flight critical communications in some airspace domains.

Another significant set of comments on the Phase I results was received from the ICAO Aeronautical Communication Panel (ACP) at the working group of the whole meeting in June 2005. Feedback to the study team on the evaluation process and criteria from the ICAO ACP indicated the original scope of the FCS was too broad. Rather than specifying a technology that would meet all of the ATM communications requirements (including voice and data), it was recommended that the technology investigation focus on a data-only solution, keeping in mind that a future system would augment existing systems, not immediately replace them. Furthermore, the ACP indicated that the genesis of the original evaluation criteria (Phase I study criteria) was unclear. The panel asked that a set of evaluation criteria directly traceable to the COCR document be developed for the FRS, and that the technology screening process be repeated (ref. 2).

All of the feedback influenced the direction of the study, helping to identify focus areas for indepth evaluations and tailoring the applied evaluation methodology.

1.4   Approach Introduction and Overview

As decision-making in the aeronautical environment can be complex, a structured methodology that accommodates stakeholder inputs has been defined and applied in the FCS. Specifically, a six-step methodology, as shown in figure 3, was defined and followed. The activities included in the methodology have been performed in three study phases, Technology Prescreening (Phase I), Technology Screening and Indepth Studies (Phase II), and Additional Technologies and Investigations for Provision of Future Aeronautical Communications (Phase III).

The first set of activities in the defined evaluation process (steps 1A and 1B) included derivation of evaluation criteria and metrics. Addressing stakeholder direction, a structured analysis of the COCR was undertaken to ensure traceability of criteria to requirements. This structured analysis, along with consideration of ICAO recommendations for future communication systems captured in consensus documentation, derived technical and viability evaluation criteria.

Step 2 is to identify the most promising technology candidates. Over 50 technologies were considered in the technology screening process. The inventory included technologies collected through requests for information from NASA to industry; EUROCONTROL inputs from European manufacturers; and ICAO ACP WG–C member state inputs and represented technologies defined for current and planned commercial applications as well as standards and prototypes developed specifically for aviation. A screening process applied a small set of key technical and viability evaluation criteria at a high level. The output of this work was a subset of most promising technologies that would be subject to indepth analysis and further consideration in the future aeronautical communication infrastructure.

The remaining steps in the evaluation process (steps 3 through 6) contribute to detailed assessment of the most promising candidate technologies. A concept of how the technology would be applied to the aeronautical environment described in the COCR was defined, followed by evaluation of a technology to the full complement of evaluation criteria. Supporting these steps was indepth analysis of the considered technologies. The process concludes with consideration of the relative importance of criteria and the use of this information to identify the best performing technology.

1.5   Supporting Assessments

A considerable number of indepth assessments were performed to support the technology evaluation process and to gain a better understanding of the applicability of the most promising technologies to the future aeronautical communication environment. Indepth studies were conducted as part of the FCS Phase II and Phase III study efforts. A full set of the indepth analyses is provided below in table I. Also indicated is a reference which identifies where the full study is documented.

TABLE I.—FCS TECHNOLOGY INVESTIGATION INDEPTH STUDIES

Topic no.	Indepth study topic	Location of study documentation (objectives, methodology, results)
1	L-band Air/Ground Communication Channel Characterization	FCS Phase II interim report ("Identification of Technologies for Provision of Future Aeronautical Communications," NASA Report NASA/CR—2006-214451, ITT Industries, July 2006), Section E.1.1
2	Project-34/Telecommunication Industry Association (TIA) 902 Series Standards (P34/TIA–902) Technology Performance Assessment	FCS Phase II interim report ("Identification of Technologies for Provision of Future Aeronautical Communications," NASA Report NASA/CR—2006-214451, ITT Industries, July 2006), Section E.1.2 and E.1.4
3	P34/TIA–902 Technology Intellectual Property Assessment	Section 4
4	L-band Digital Link (LDL) Technology Performance Assessment	FCS Phase II interim report ("Identification of Technologies for Provision of Future Aeronautical Communications," NASA Report NASA/CR—2006-214451, ITT Industries, July 2006), Section E.1.3 and E.1.4
5	Wideband Code Division Multiple Access (WCDMA) Functional Assessment	Section 3
6	L-band Technology Cost Assessment for Ground Infrastructure	FCS Phase II interim report ("Identification of Technologies for Provision of Future Aeronautical Communications," NASA Report NASA/CR—2006-214451, ITT Industries, July 2006), Section E.1.8
7	L-band Interference Testing	Section 2
8	Satellite Technology Availability Performance	FCS Phase II interim report ("Identification of Technologies for Provision of Future Aeronautical Communications," NASA Report NASA/CR—2006-214451, ITT Industries, July 2006), Section E.2
9	IEEE 802.16e Performance Assessment in Aeronautical C-band Channel	FCS Phase II interim report ("Identification of Technologies for Provision of Future Aeronautical Communications," NASA Report NASA/CR—2006-214451, ITT Industries, July 2006), Section E.3

2.0 L-Band Interference Measurements

This section of the report describes one of the indepth analyses conducted during FCS Phase III. Four major sections document the work: background, methodology, results, and conclusions. The background section describes the motivation, objective, and approach for this study and provides a brief overview of the distance measuring equipment (DME) system. The methodology section describes the test setup, test equipment, test cases, and waveform development process. The results section presents plots of the measurement data and explains how the data should be interpreted. The conclusions section describes how the data should be applied towards the technology evaluations and also suggests additional tests that would build upon this work and improve understanding of the L-band interference environment.

2.1　Background

ICAO is considering the use of L-band (960 to 1024 megahertz (MHz)) to employ the next generation aeronautical communication system. In order to assess the viability of proposed communication systems in this frequency band, the interference potential from FRS candidates to systems already utilizing this spectrum must be characterized. An interference measurements campaign was defined for Phase III of the FCS to better our understanding of the interference environment in L-band. ITT, with the help of the Ohio University Avionics Engineering Department, conducted interference measurements against DME for three FRS candidate technologies: WCDMA, LDL, and P34.

2.1.1　Motivation

The DME system is a vital navigation tool used to ensure flight safety. Before any new communication systems can be allowed to share spectrum with DME, a compatibility analysis must be performed to assess potential degradation of DME system performance. A compatibility analysis of a FRS with DME in L-band involves many potential interference scenarios, including cosite (onboard the aircraft), air-to-air, air-to-ground/ground-to-air, and ground-to-ground (ref. 4). These various scenarios are illustrated in figure 4.

For this interference measurement study, the cosite interference scenario is emulated. The cosite interference scenario serves as a guide for specifying the bounds of interference power used when taking measurements. The signal levels used for interference measurements should encompass the power ranges that will be seen in practice, and these levels are greatest in the cosite scenario. This is not to say that ground stations do not also transmit high power signals, but ground stations are generally separated by large distances to mitigate interference effects. Antennae onboard the aircraft do not have this liberty. Figure 5 illustrates the cosite scenario.

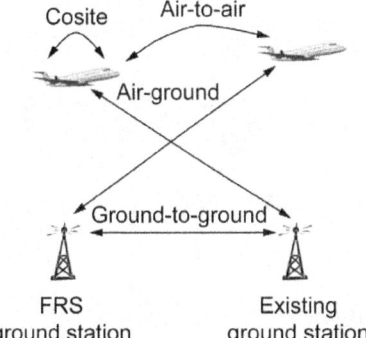

Figure 4.—Interference scenarios for compatibility analysis.

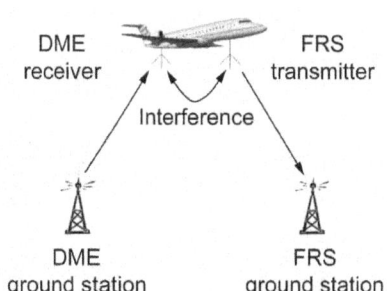

Figure 5.—Cosite interference scenario.

2.1.2 Objective

The objective of this study is to determine the extent and nature of interference to DME airborne receivers by FRS candidate technology waveforms. This study focuses only on the susceptibility of interference to existing airborne avionics. This study does not focus on interference from DME to FRS airborne avionics, as L-band FRS receivers do not yet exist.

The three FRS candidate technologies selected for this interference study are WCDMA, LDL, and P34. All three technologies scored well in the prescreening and screening stages conducted during Phases I and II of the FCS (ref. 5). WCDMA scored highest out of the cellular telephony technology family. LDL scored highest out of the custom narrowband technology family. P34/TIA–902 scored highest out of the public safety radio technology family. In addition to representing their respective technology families, WCDMA, LDL, and P34, also use fundamentally different waveform structures: ultrawideband, narrowband, and orthogonal frequency division multiplexing (OFDM), respectively.

2.1.3 Approach

The overall approach for the interference measurements study consists of six interrelated tasks. Task 1 consists of generating a test plan that contains test procedures for each test case in the study. Task 2 consists of specifying the interference sources to be used for this study. Task 3 consists of procuring test equipment to emulate FRS transmitters. Task 4 consists of conducting bench tests against the DME receivers. Task 5 consists of analyzing and reducing the measurement data. The sixth task consists of documenting the study and presenting results in a report. The flow of these tasks is illustrated in figure 6.

2.1.4 DME System Overview

The DME system is a navigational aid that provides slant range distance to aircraft. The system consists of airborne equipment and ground-based equipment, which are called interrogators and transponders, respectively. The interrogator transmits paired pulses to the transponder which replies with its own paired pulse message. The interrogator measures the time elapsed between its own transmission and reception of the transponder's response, which is then used to calculate the slant range distance. Figure 7 illustrates this concept.

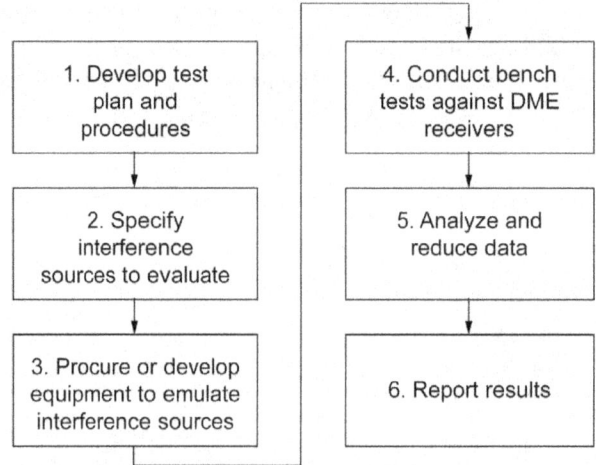

Figure 6.—Work flowchart for interference measurements study.

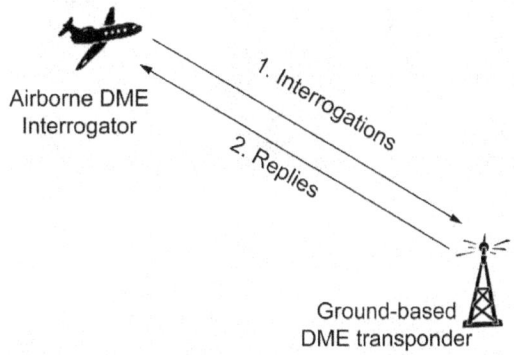

Figure 7.—DME concept of operation.

The DME system operates in the frequency band 960 to 1215 MHz. The interrogation and reply frequencies are always offset by 63 MHz. The entire band allows 126 channels for interrogation and 126 channels for transponder replies. Interrogations are sent on frequencies 1025 to 1150 MHz. Replies from the transponders are sent on frequencies 962 to 1024 MHz and 1151 to 1213 MHz. DME channels are spaced 1 MHz apart on center, with bandwidths of 0.5 MHz.

2.2   Methodology

The basic methodology for characterizing interference susceptibility is to observe the response of the DME interrogator in the presence of the interfering signals. The interfering waveforms are injected into the system at various frequency separations, and the signal levels are incrementally adjusted in order to determine the power thresholds that induce a standard response from the device under test (DUT).

For DME, the standard response is described by two metrics: acquire stable operating point (ASOP) and break stable operating point (BSOP). ASOP is defined as the maximum interfering signal level where the DME interrogator consistently locks on and provides bearing, ident, and range information. BSOP is defined as the minimum interfering signal level that causes the DME interrogator to lose lock by failing to provide bearing, ident, or range information. These definitions are reiterated below in table II.

TABLE II.—ASOP AND BSOP DEFINITIONS (ref. 6)

DME standard response metrics	Definition
ASOP	Desired signal level at which bearing, ident, and range lock on consistently
BSOP	Desired signal level at which either bearing, ident, or range breaks lock

2.2.1   Test Setup and Equipment

The test setup for the interference measurements is illustrated in figure 8. Interrogations are sent from the DME interrogator to the DME test set. The DME test set replies to interrogations with a DME reply signal. The DME reply signal is coupled with the interference signal, and the DUT is observed. The interference signal undergoes bandpass filtering before it is coupled with the DME reply signal. This reduces transmitter broadband noise that is produced by the vector signal generator. The spectrum analyzer allows test engineers to observe spectrums of the signals in the system. Detailed descriptions of these devices follow the test setup diagram.

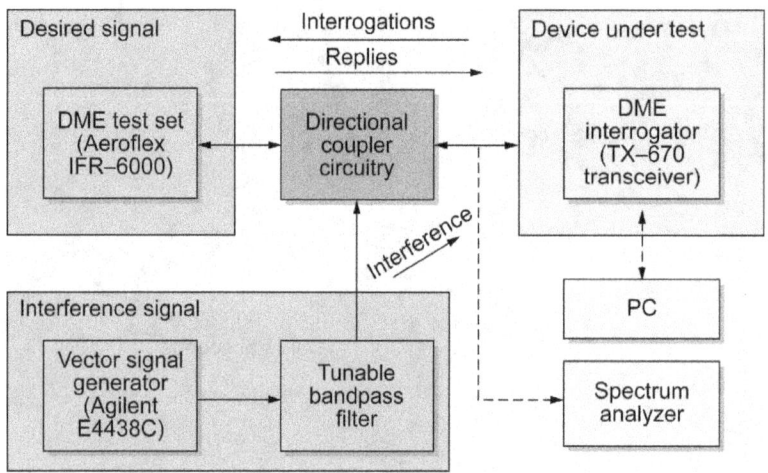

Figure 8.—Test setup for DME interference measurements.

The DUT for this study is the DME interrogator. The make and model that was tested is the BF Goodrich Flights Systems TX–670 DME Transceiver (BF Goodrich). Ideally, models from various manufacturers would be studied to develop a performance range, but testing of additional interrogators was not possible in the context of this study. The TX–670 unit is a cabin-class DME interrogator. This particular DME interrogator has been used in a variety of FAA-sponsored DME tests conducted by Ohio University including design approval tests of DME ground stations.

The ground-based transponder is emulated by the Aeroflex IFR–6000 Ramp Test Set (Aeroflex). This multimode test is designed to test avionics from various systems, including DME. The IFR–6000 both receives and replies to interrogations, while providing insight into the current state of the interrogator through a real-time display screen. Several useful status parameters such as DME reply signal level and pulse repetition frequency (PRF) are displayed and updated during testing.

The interference source is emulated by an Agilent E4438C Vector Signal Generator (Agilent). This device allows users to create waveform files in software and then upload those waveform files to the signal generator for playback. Four options for the vector signal generator were purchased to facilitate testing. Option 005 includes a 6-GB internal hard drive for storing waveform files. Option 502 offers the ability to transmit up to 2 GHz. Option 602 includes baseband signal generation capabilities. Option 400 includes the WCDMA personality, which is described in more detail in Section 2.2.3.

The filter used in the test setup is a Trilithic tunable bandpass filter (model number: 5VF–750/1500–1–50–KK) (Trilithic). Results from the first pass of interference measurements indicated the presence of transmitter broadband noise in the interference signals. The broadband noise was indistinguishable from the noise floor of the spectrum analyzer during waveform validation. Discussions with the equipment manufacturer verified that the noise floor is attributable to excess quantization noise in the vector signal generator, and the only way to reduce the noise floor was with additional filtering. The additional filtering reduces the noise floor, which results in a more realistic spectrum that one would expect from actual transmitter implementation. Figure 9 shows the filter response to a sine wave sweep when centered at the chosen test frequency, 966 MHz. The filter response shows a passband of about 12 MHz and provides at least 37 dB of attenuation only 10 MHz away from the passband. The insertion loss of the filter and associated cables totals 2.5 dB.

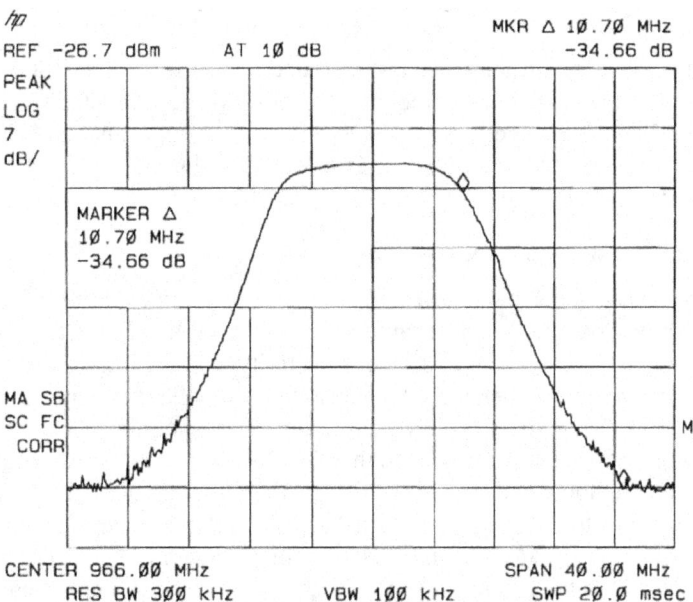

Figure 9.—Filter response of sine wave sweep.

The personal computer interfaces with the interrogator via a software program. The program collects and displays status data during testing. This data, which includes bearing, ident, and range, is used to determine if the interrogator has acquired or broken lock. The software stores the data from each trial into a unique file for postprocessing.

The make and model of the spectrum analyzer is Hewlett Packard 8591E (Hewlett Packard). The spectrum analyzer is used to capture spectrum plots of the signals in the system.

2.2.2   Test Cases

Four major test cases comprise the interference measurements study. The test procedures for each of these test cases were derived from measurement procedures outlined in MIL–STD–449D (ref. 7). Below are descriptions and general procedures for each test case.

2.2.2.1   Receiver Sensitivity Test

The purpose of the receiver sensitivity test is to determine power thresholds of the DME reply signal that induce a standard response from the DME interrogator in the absence of an interference signal. To start, the DME reply signal level is set very low (~ –100 dBm) so that the DME interrogator cannot achieve lock. Using the DME test set, the DME reply signal level is incrementally increased until the DME interrogator achieves lock. Next, the DME reply signal level is incrementally decreased until the DME interrogator breaks lock. This test characterizes the hysteresis of acquiring and breaking lock of the receiver.

2.2.2.2   Transmission Emission Spectrum Test

The purpose of the transmission emission test is to capture the power vs. frequency characteristics of the interference waveforms. During this test the interference source is connected directly to the spectrum analyzer, and the frequency spectrum is saved to disk.

2.2.2.3   Co-Channel Interference Test

The purpose of the co-channel interference test is to determine ASOP and BSOP for the interfering waveforms when they are tuned to the same frequency as the DME interrogator. To start, the DME reply signal level is set to 3 dB above receiver sensitivity, and the interfering signal level is incrementally adjusted until ASOP and BSOP are found. The test is repeated for higher DME reply levels (in 10-dB increments) until a DME reply level close to –50 dBm is reached.

2.2.2.4   Adjacent Channel Interference Test

The purpose of the adjacent channel interference test is to determine ASOP and BSOP for the interfering waveforms at various frequency separations from the DME tuned frequency. To start, the DME reply signal level is set to 3 dB above receiver sensitivity, and the interfering signal level is set as high as possible. The vector signal generator used during this study has a maximum power output of 20 dBm (9.57 dBm after insertion losses). With the interference power maxed out, the frequency of the interference source is tuned far enough away such that the DME interrogator can acquire lock. Next, the frequency separation is incrementally decreased until the DME interrogator breaks lock. The frequency separation where the interrogator loses lock is defined as Δf. Next, ASOP and BSOP are determined at frequency separations of Δf, $\frac{1}{2}\Delta f$, $\frac{1}{4}\Delta f$, etc. until the frequency separation is within the 3 dB bandwidth of the DME signal. The data from the co-channel and adjacent channel tests are plotted as sets of ASOP and BSOP curves for various DME reply levels.

2.2.3   Waveform Development

L-band transmitters do not exist for the three FRS candidate technologies tested, so these waveforms must be emulated by test equipment. The vector signal generator procured offers various solutions for developing these waveforms. The waveform development process for each of the FRS candidate technologies tested in this study is described below.

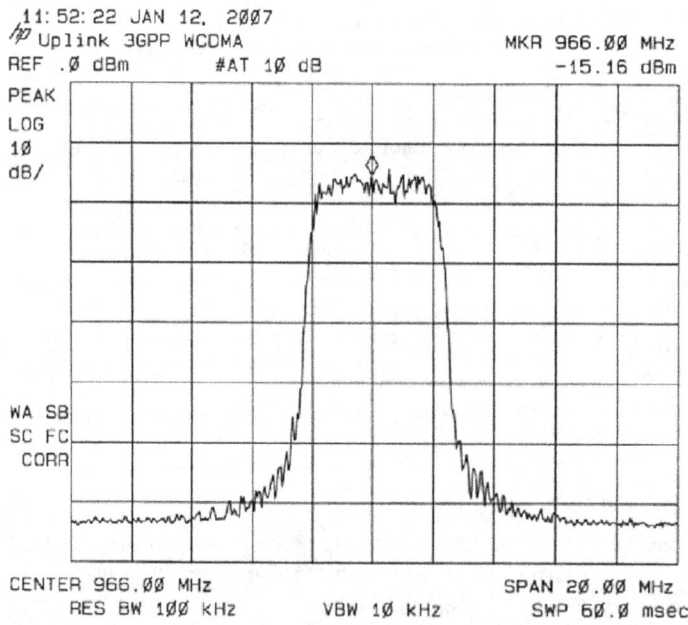

Figure 10.—WCDMA transmission emission spectrum.

2.2.3.1   WCDMA Waveform Development

Option 400 for the E4438C vector signal generator is the WCDMA personality. This personality is a preloaded waveform used by industry to test user and base station equipment for the WCDMA standard. No additional tailoring of this waveform was required. Figure 10 shows the emission spectrum of the WCDMA waveform whose bandwidth is 5 MHz.

2.2.3.2   LDL Waveform Development

No waveform personality exists for LDL so the LDL waveforms are developed using MATLAB and Simulink.[1] A transmitter model for LDL was created in Simulink in accordance with the published documentation (ref. 8). The model produces a complex baseband representation of the waveform. The waveform is formatted so that it can be uploaded to the vector signal generator. Agilent offers a free MATLAB library that transfers formatted waveforms to the signal generator. Once transferred, the waveform segments can be looped to play for as long as is required. Figure 11 shows the Simulink model used to generate the LDL waveforms.

Two versions of the LDL waveform are modeled for the interference measurements study. The first is a gated LDL waveform. The gating structure was modeled after the "Simple LDL Time Slot Structure" described in the June 2005 paper "An L-band Digital Communications Link Concept for Air Traffic Control" by Dr. Warren J. Wilson. The time slot is 30 µs in duration and contains 1875 bits. The time slot is composed of four segments in the following order: a 72-bit M-burst, a 589-bit guard period, a 624-bit V/D-burst, and a 590-bit guard period. A mask that uses this structure is applied to the waveform to achieve the gating mechanism. Figure 12 shows a plot of the mask where the abscissa is scaled to waveform samples (187 500 samples total).

The overall duty cycle of the gated LDL waveform is 37.12 percent. The second LDL waveform modeled is a continuous waveform whose duty cycle is 100 percent. Figure 13 shows the transmission emission spectrum of the continuous LDL waveform.

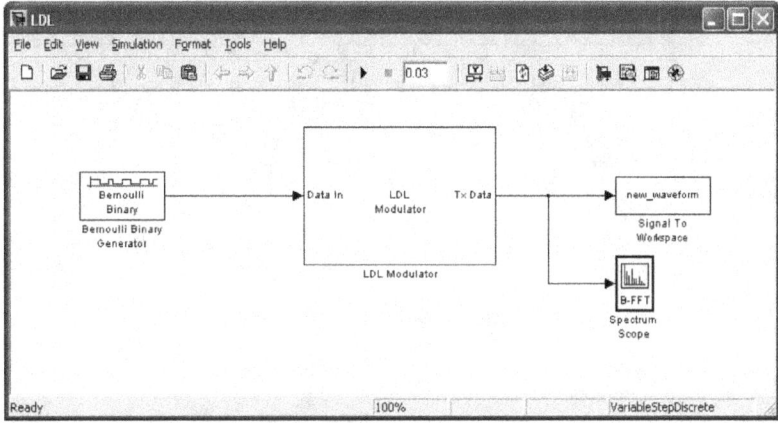

Figure 11.—LDL transmitter Simulink model.

---

[1] MATLAB and Simulink are registered trademarks of The Mathworks.

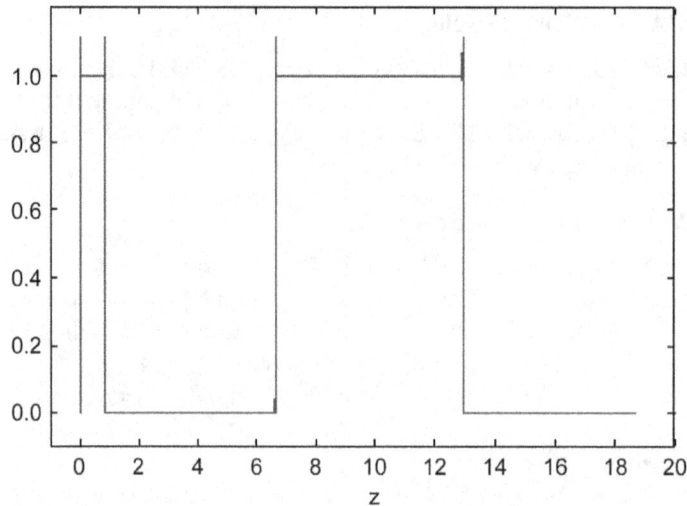

Figure 12.—Mask for gating LDL waveform.

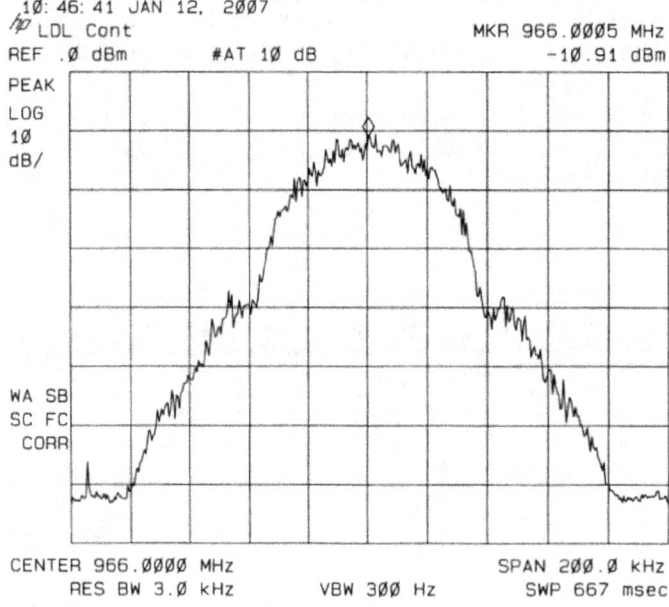

Figure 13.—LDL transmission emission spectrum.

2.2.3.3   P34/TIA–902 Waveform Development

For this assessment, the P34/TIA–902 ground station waveform (continuous) was modeled. The aircraft station waveform, which is a gated waveform, was not modeled in this assessment. Therefore, the results of the P34/TIA–902 measurements can be considered an upper bound on interference or can be used for assessment of the intersite interference scenario. The P34/TIA–902 waveform development process was the same as used for LDL, with the exception of the gating mechanism. Figure 14 shows the transmitter model for P34.

Each mode of P34/TIA–902 (50-, 100-, and 150-kHz) has its own Simulink model, but only the 50-kHz is shown. All of the P34/TIA–902 waveforms are modeled as continuous transmissions (100 percent duty cycle). Figure 15 shows the transmission emission spectrum of the 50-kHz mode of P34.

Figure 16 shows the transmission emission spectrum of the 100-kHz mode of P34.

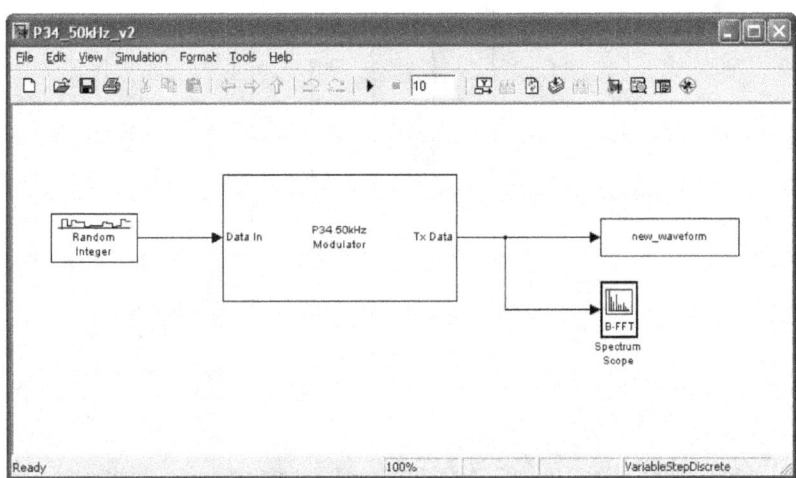

Figure 14.—P34/TIA–902 50-kHz transmitter Simulink model.

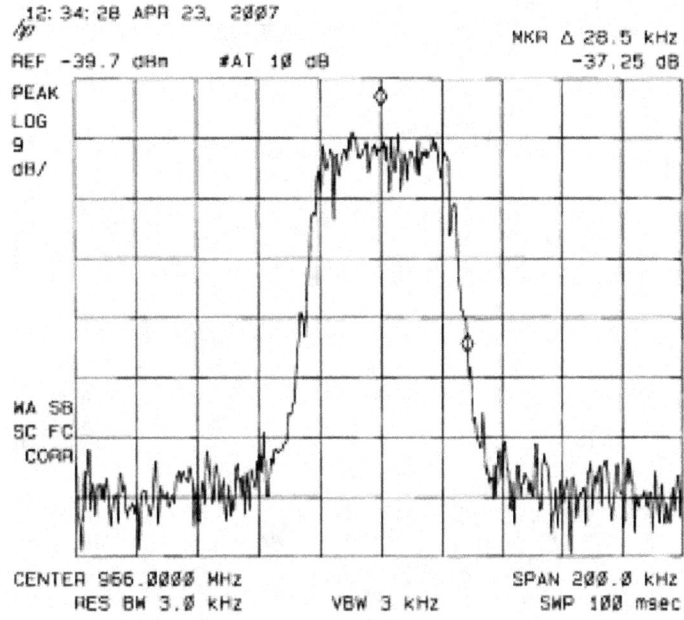

Figure 15.—P34/TIA–902 (50 kHz) transmission emission spectrum.

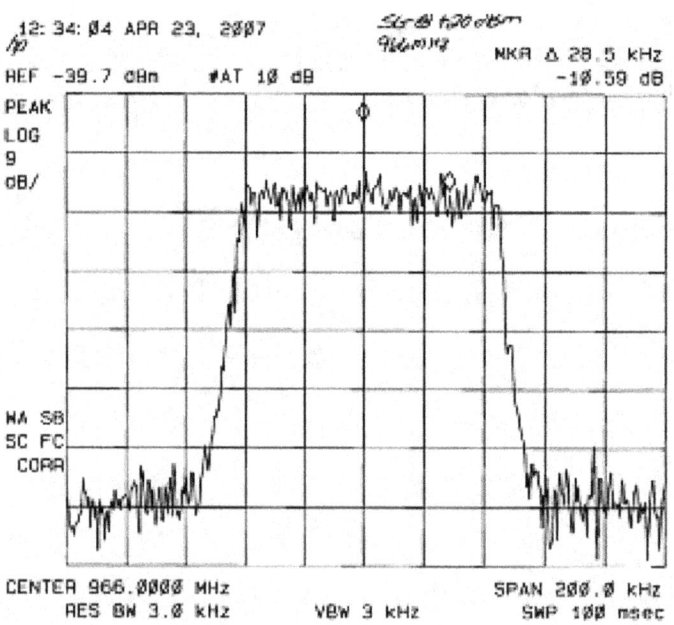

Figure 16.—P34/TIA–902 (100 kHz) transmission emission spectrum.

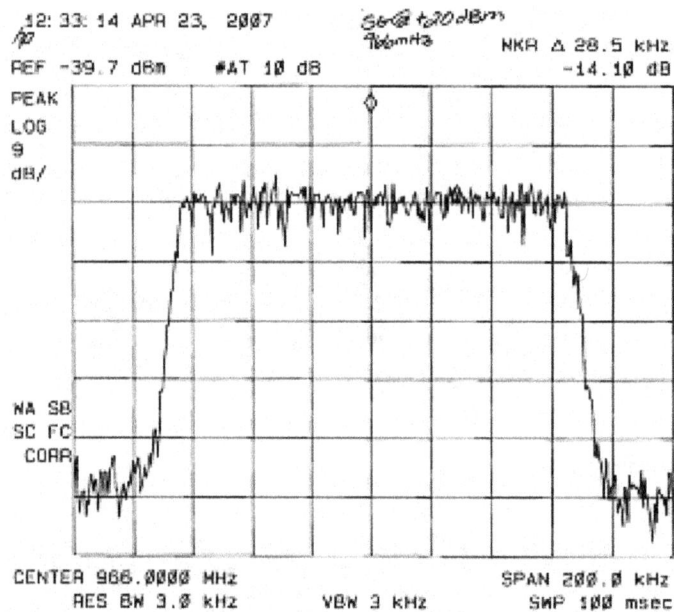

Figure 17.—P34/TIA–902 (150 kHz) transmission emission spectrum.

Figure 17 shows the transmission emission spectrum of the 150-kHz mode of P34.

2.3   Measurement Results

This section is organized by waveform and begins with a discussion of the expected undesired power levels at the DME receiver. Next, ASOP and BSOP charts show the actual measurements. The charts are followed by a discussion of how they should be interpreted. In general, the ASOP and BSOP charts in this section contain data from the both filtered and unfiltered trials. Specifically, any data for frequencies greater than or equal to 980 MHz or any data for frequencies less than or equal to 952 MHz use the filter. Likewise, any data from frequencies greater than 952 MHz, but less than 980 MHz do not use the filter. Complete datasets for the unfiltered measurements can be found in appendix A.

In an attempt to minimize confusion when reviewing the results, some questions been addressed here.

- **Which plot is the author talking about, ASOP or BSOP?** The discussions that follow the charts speak to the ASOP charts. The purpose of this interference study is to determine the maximum interference levels at which the DME can acquire lock, not break lock.

- **Why are the plots not left-right symmetrical about 966 MHz?** The DME receive band spans 960 to 1215 MHz and the selectivity of the DME interrogator attenuates the signals outside of this band. Often times data from frequencies less than 966 MHz appears to be shifted upwards. The test frequency, 966 MHz, was chosen to pick up this selectivity.

- **Why is there no data below 966 MHz?** In order to have enough time to collect at least some data for all of the waveforms, some of the datasets do not have data from below 966 MHz. The most important data is in the DME band: 960 to 1215 MHz.

- **As the frequency separation increases in the ASOP plots, why do data points seem to eventually converge to 9.57 dBm? Also, why does it look like data is missing from some of the BSOP plots?** These two questions describe the same scenario. For some of the interference waveforms, especially at large frequency separations, the maximum power of the signal generator is not enough to cause the DME to break lock. In these instances, 9.57 dBm is recorded for ASOP since the DME can acquire at maximum power. Likewise, the corresponding point in the BSOP plot spears to be missing because an actual BSOP could not be determined. In reality the ASOP value might be higher than 9.57 dBm.

- **Why is there a gap in the data between 952 and 980 MHz?** If no data exists between 952 and 980 MHz, then only filtered trials were conducted for that waveform. If data does exist in this range, then that data is from unfiltered trails.

2.3.1   WCDMA/DME Interference Measurements

Link budgets for a representative FRS WCDMA implementation indicate a transmit powers on the order of 33 dBm (ref. 9). Typical values of isolation for antennae on the same side of the aircraft range are 25 to 35 dB. Using an optimistic value of 35 dB for isolation, the resultant undesired power at the DME receiver is –2 dBm. This value is important when interpreting the measurements shown in figure 18. Note that the cosite power level is shown on the figure; the DME received signal level should be above this line for DME operations (ASOP). For reference (as the WCDMA ground station transmitter has a similar waveform to what has been modeled for an aircraft transmitter, and assuming an aircraft 1000 ft above a ground transmitter), an intersite interference reference power level is also noted in figure 18.

The measurements show that the DME interrogator breaks lock for WCDMA signal levels less than what is expected on the aircraft. This phenomena occurs even at large frequency separations. It is not until the interference signal is tuned 50+ MHz away from the DME where all of the ASOP curves are above the expected power threshold of –2 dBm. The WCDMA waveform tested in this study has a 100 percent duty cycle.

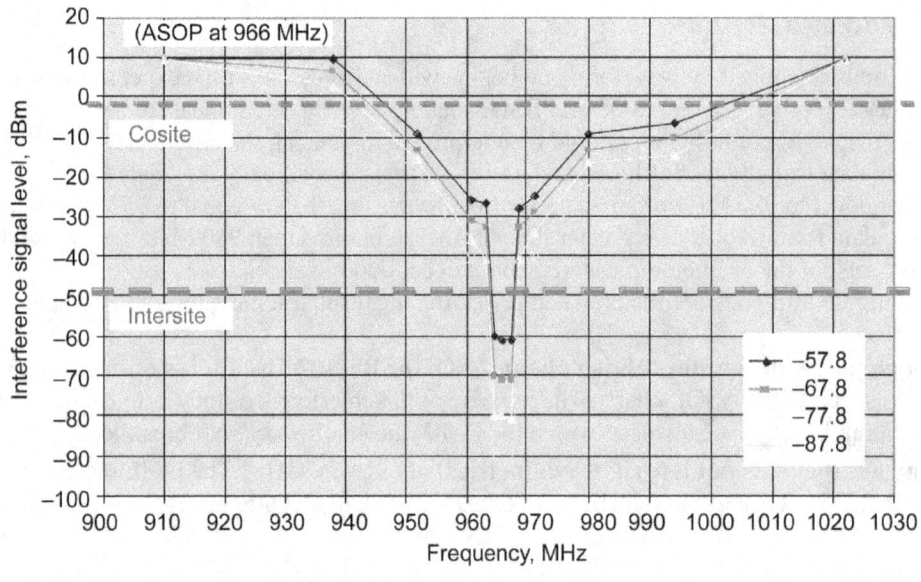

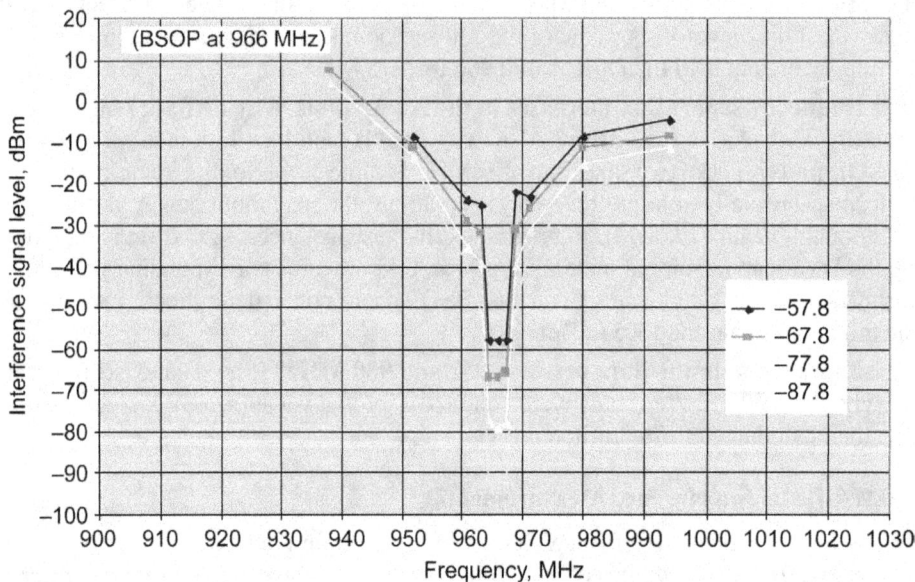

Figure 18.—WCDMA/DME interference measurements.

2.3.2 LDL/DME Interference Measurements

Link budgets for LDL in a FRS implementation indicate transmit power levels on the order of 40 dBm. Again, using an optimistic value of –35 dB for antenna isolation, the resultant undesired power at the DME receiver is 5 dBm. Figure 19 shows the results for the nongated LDL waveform.

The measurements show that the DME interrogator breaks lock for nongated LDL signal levels less than what is expected on the aircraft. This phenomenon occurs even at large frequency separations. It is not until the interference signal is tuned 50+ MHz away from the DME where all of the ASOP curves are above the expected power threshold of 5 dBm.

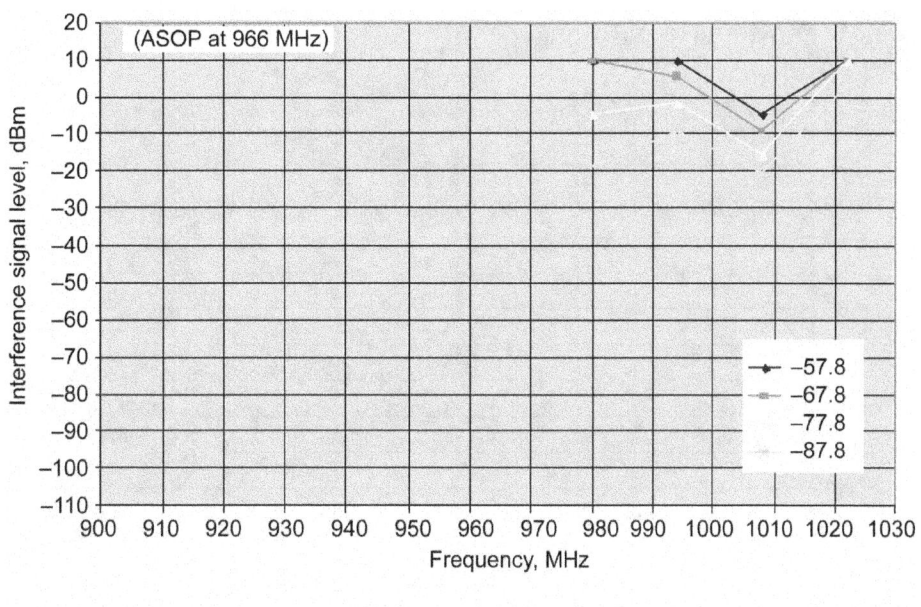

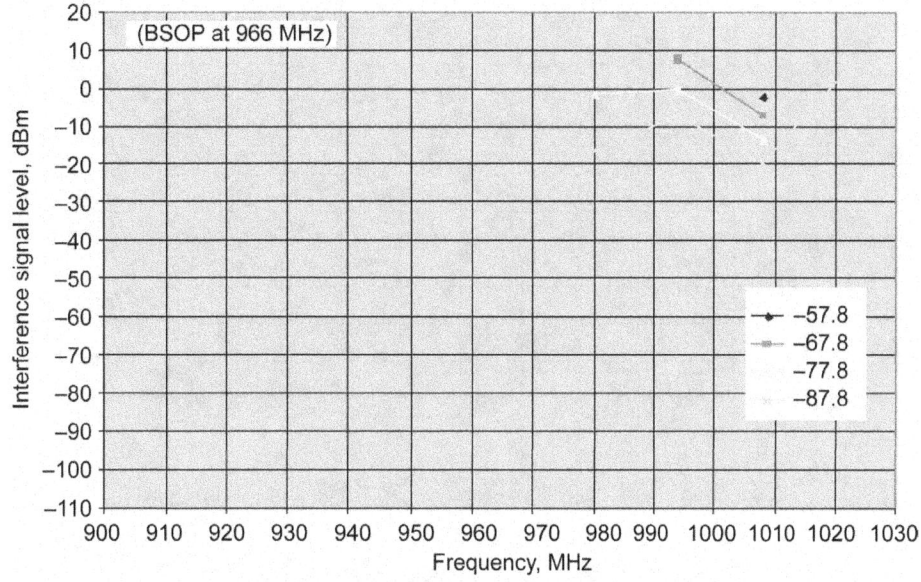

Figure 19.—Nongated LDL/DME interference measurements.

LDL uses very high frequency digital link 3 standard for protocols layers above the physical layer, including a gated/time division multiple access (TDMA) framing structure. This suggests that a duty cycle far less than 100 percent might be expected in practice. A gated version of the waveform was also tested; figure 20 shows the results for the gated LDL waveform. As in the WCDMA results, the graph below includes the cosite interference reference line and the intersite interference reference line (assuming an aircraft 1000 ft above a ground transmitter). The intersite reference is applicable as the LDL ground radio may have similar waveform and/or duty cycle as the aircraft transmitter.

The gated version did not cause the DME to break lock for filtered trials and the DME was able to acquire lock under maximum interference conditions (test frequencies off-tuned by 10+ dB). Almost all of the filtered ASOP values were recorded as 9.57 dBm, inferring that the ASOP values are actually higher than 9.57 dBm. The effect of gating makes the DME much more tolerant of this waveform. This

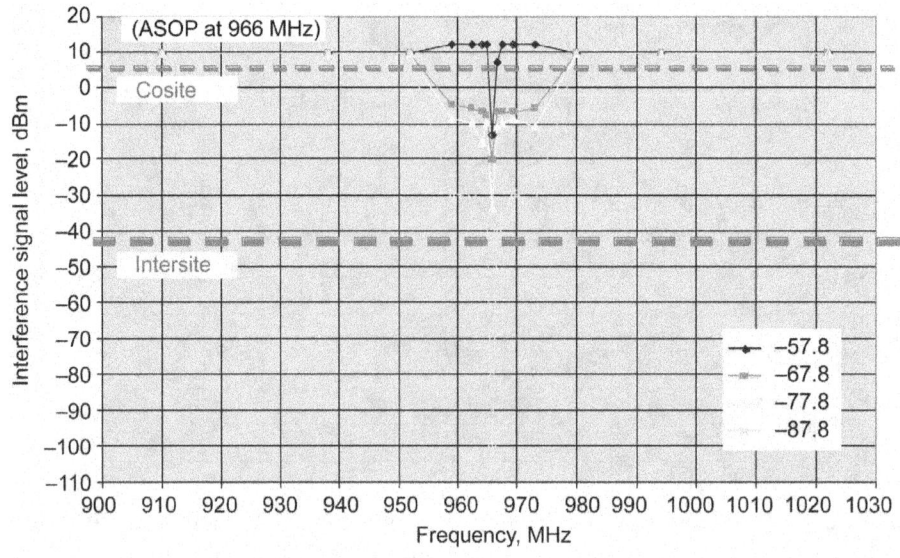

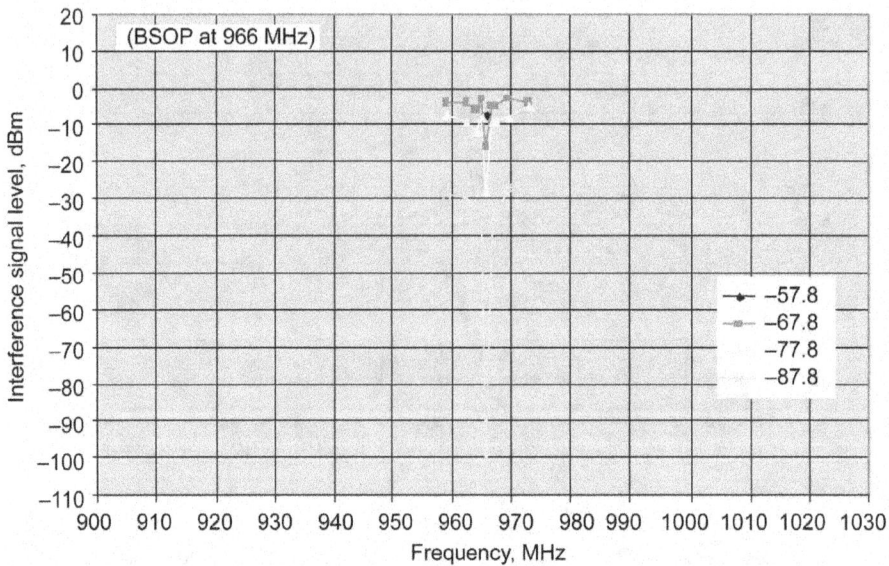

Figure 20.—Gated LDL/DME interference measurements.

suggests that interference susceptibility is related to the average power of the interferer. This was the only gated waveform tested during this study, and the results for it were the most positive. Filtered results near the tune frequency of the DME suggested off-tuning may still be required for LDL.

2.3.3 P34/DME Interference Measurements

Link budgets for a representative P34/TIA–902 implementation suggest a similar transmit power to LDL can be used (on the order of 40 dBm). Considering a cosite scenario (not completely applicable as the modeled waveform was for the P34/TIA–902 ground stations) and applying an optimistic value of –35 dB for antenna isolation, the resultant undesired power at the DME receiver is 5 dBm. For an intersite scenario, more applicable for this modeled waveform, assuming that an aircraft is 1000 ft above the transmitting ground radio, the path loss is on the order of 83 dB and thus the resultant undesired power at the DME receiver is –43 dBm. Figure 21 shows the results for the 50-kHz mode of P34/TIA–902 (constant transmission).

The measurements show that the DME interrogator breaks lock for the P34/TIA–902 (50 kHz) signal levels less than what is expected on the aircraft (assuming unrealistic 100 percent duty-cycle model). This phenomena occurs even at large frequency separations. It is not until the interference signal is tuned 50+ MHz away from the DME where all of the ASOP curves are above the expected power threshold of 5 dBm. These results are similar to the results from the nongated LDL waveform and some inferences can be made. Reducing the duty cycle for the LDL waveform improved the performance of the DME significantly. The duty cycle of P34/TIA–902 in practice is surely to be less than the 100 percent duty cycle tested during this study. It is recommended that additional tests be conducted to experiment with duty cycle for this waveform.

For the intersite scenario, when the interference signal is tuned approximately 1 to 2 MHz away from the DME, the ASOP curves are above the expected power threshold of 5 dBm. These results are very promising for compatibility.

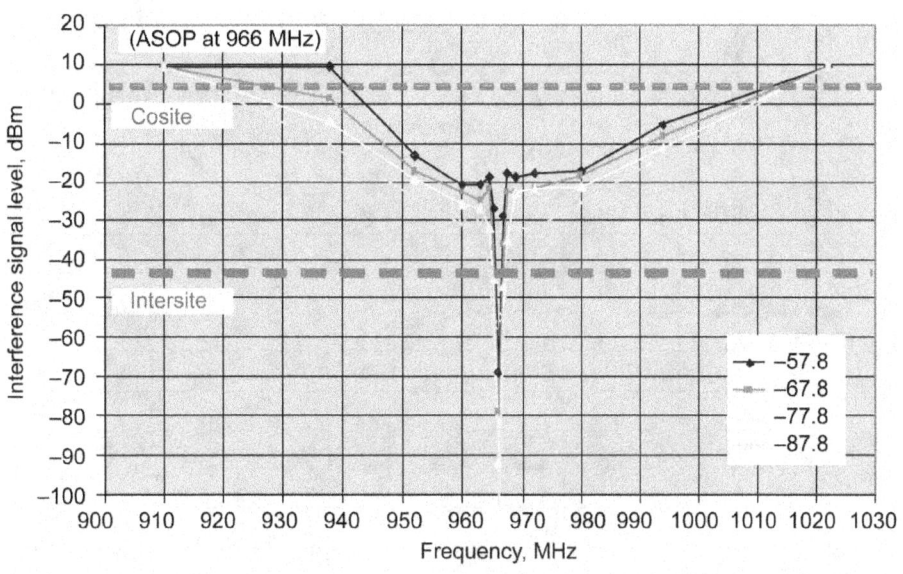

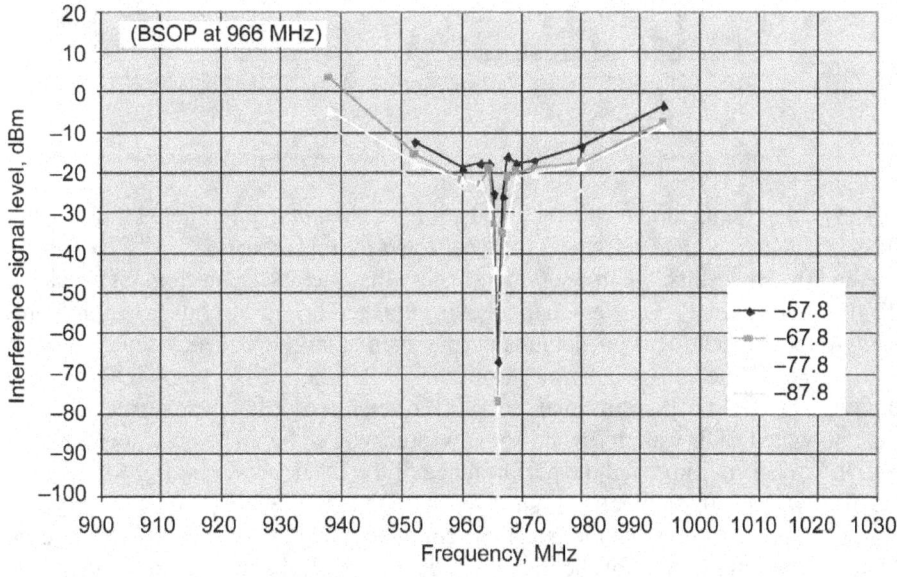

Figure 21.—P34/TIA–902 (50 kHz)/DME interference measurements.

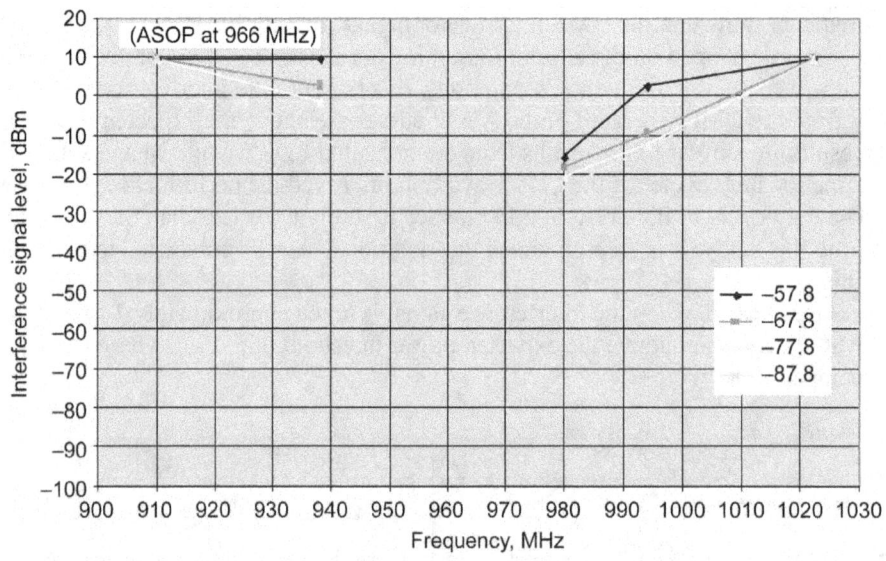

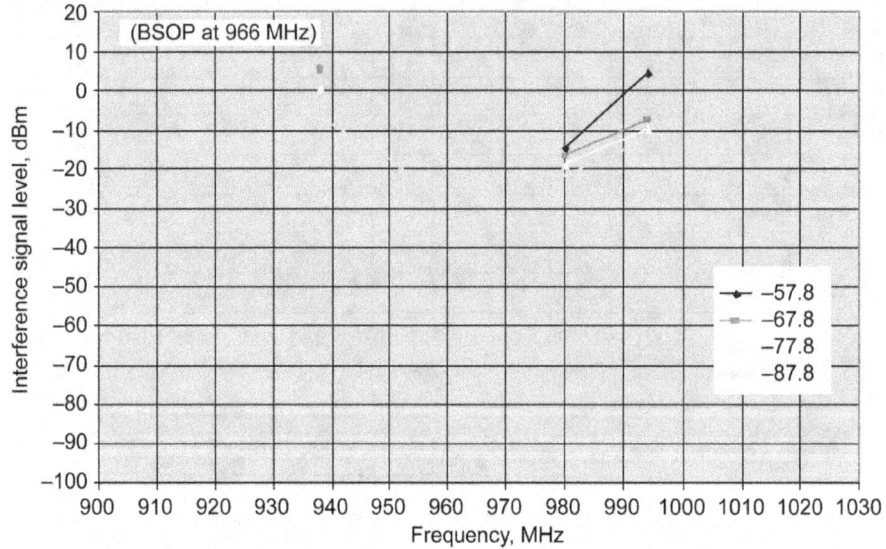

Figure 22.—P34/TIA-902 (100 kHz)/DME interference measurements.

Figure 22 shows the results for the 100-kHz mode of P34. Note that only filtered trials (for frequencies less than 952 MHz and greater than 980 MHz) were collected for this scenario.

Similar to the 50-kHz results, the measurements show that the DME interrogator breaks lock for the P34/TIA–902 (100 kHz) signal levels less than what is expected on the aircraft (assuming unrealistic 100 percent duty-cycle model). This phenomena occurs even at large frequency separations. It is not until the interference signal is tuned 50+ MHz away from the DME where all of the ASOP curves are above the expected power threshold of 5 dBm. Like the 50-kHz version of P34, these results are similar to the results from the nongated LDL waveform and some inferences can be made here, too. Reducing the duty cycle for the LDL waveform improved the performance of the DME significantly. As noted previously, the duty cycle of P34/TIA–902 in practice is surely to be less than the 100 percent duty cycle tested during this study. We recommend that more tests be conducted that experiment with duty cycle for this waveform. Note, that testing results for the intersite scenario, the fully applicable scenario for the continuous P34/TIA–902 waveform, were not conducted for the 100-kHz bandwidth. It is anticipated, however, that results similar to the 50-kHz channel would be achieved.

Figure 23 shows the results for the 150-kHz mode of P34. Note that only filtered trials (for frequencies less than 952 MHz and greater than 980 MHz) were collected for this scenario.

Once more, the measurements show that the DME interrogator breaks lock for the P34/TIA–902 (150 kHz) signal levels less than what is expected on the aircraft (assuming unrealistic 100 percent duty-cycle model). This phenomena occurs even at large frequency separations. It is not until the interference signal is tuned 50+ MHz away from the DME where all of the ASOP curves are above the expected power threshold of 5 dBm. Like the 50- and 100-kHz versions of P34, these results are similar to the results from the nongated LDL waveform and the inferences as noted earlier can be made here, too. Note, that testing results for the intersite scenario, the fully applicable scenario for the continuous P34/TIA–902 waveform, were not conducted for the 150-kHz bandwidth.

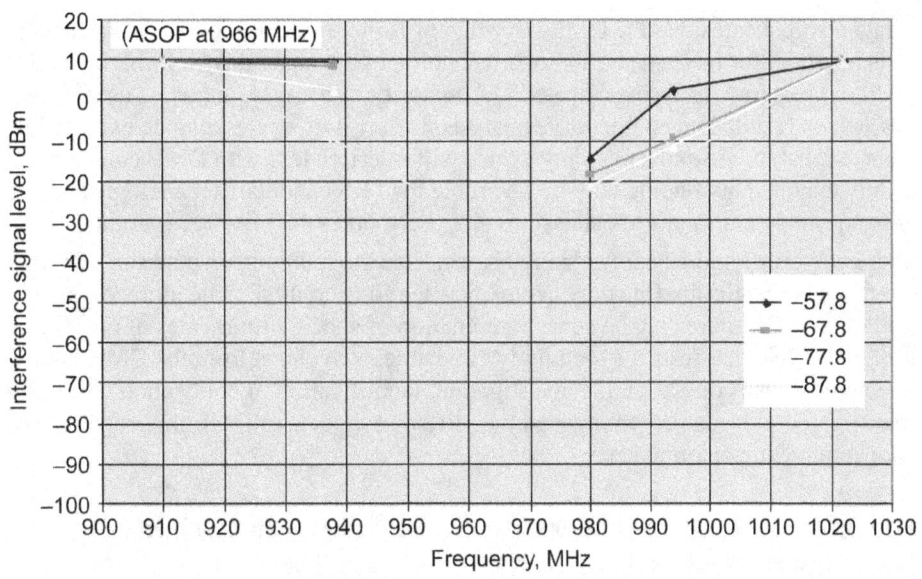

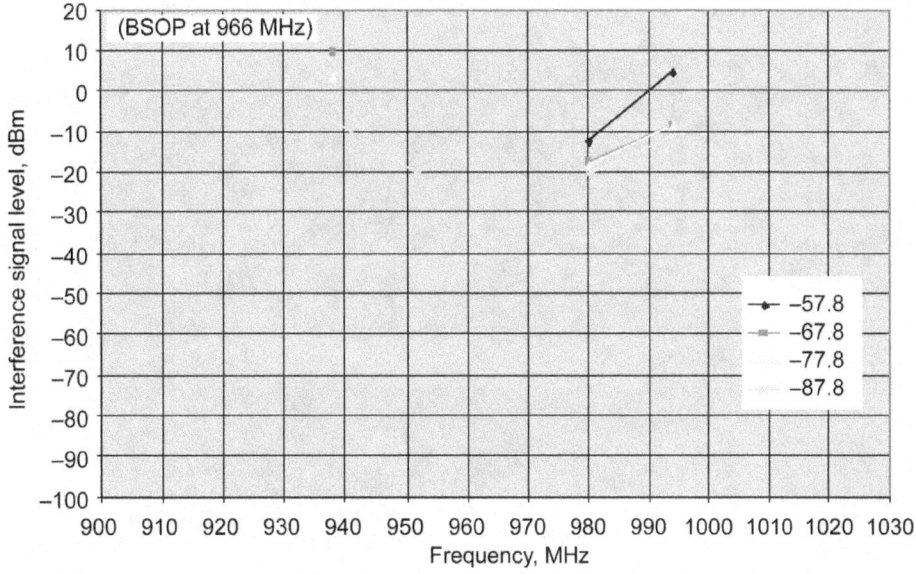

Figure 23.—P34/TIA–902 (150 kHz)/DME interference measurements.

2.4   Conclusions and Future Work

This study provides an improved understanding of the interference environment associated with DME. While limitations in the test equipment do not provide enough information to address specific channelization techniques for FRS candidates, the interference measurements do suggest that the addition of an FRS in the DME band may be feasible. Three major findings of the interference measurements study are as follows:

- The power levels expected from continuously transmitting FRS equipment onboard the aircraft may be sufficiently high as to cause desensitization in the DME interrogator. This phenomenon was evident for all of the FRS candidates even at large frequency separations for the DME that was tested. This finding is not favorable for FRS candidate technologies whose concept of use assumes continuous transmissions (e.g., WCDMA).

- The data also indicates that the DME interrogator is more tolerant to gated transmissions (i.e., there is potential for implementation of a technology with a gated waveform; but offset channels may still be required (to be investigated)). A majority of the measurements used 100 percent duty cycles, which results in a conservative analysis. Lesser duty cycles may be expected in practice. It is expected that low-to-moderate duty cycles will interfere less with DME compared to FRSs with high duty cycles. This finding may be favorable for FRS candidate technologies whose concept of use assumes noncontinuous transmissions (i.e., LDL and P34/TIA–902 (partial)).

- Further analysis should be conducted to characterize the relationship between FRS duty-cycle and interference susceptibility (the duty cycle investigation should include more variables than just overall duty cycle; there may be some combination of specific timescales of on/off pulses and overall duty cycle that results a seemingly "invisible" waveform from the DME interrogator's perspective); in the context of this investigation, identification of collocation constraints can also be investigated. It is also recommended that different models of DME interrogators be tested to provide a range of performance.

The interference measurements, in conjunction with the concept of use defined for each candidate FRS technology, were used to support the detailed technology evaluations in this study. It is recommended that prototype FRS transmitters be developed to facilitate additional technology evaluations and/or modifications and to help define specific channelization techniques.

3.0 WCDMA Functional Assessment

Detailed technology analyses conducted as part of the FCS Phase III effort focused on two of the candidate FRS solutions (to complement previous indepth assessments). Specifically, one of the technologies and analysis areas addressed is to perform functional analysis of WCDMA services, create mapping of COCR services to WCDMA services/functions, and trace protocol interactions.

WCDMA is a direct spread, wideband frequency division, duplex code division multiple access (CDMA) standard developed by 3GPP (Third Generation Partnership Project). WCDMA was selected as the air interface for the Universal Mobile Telecommunications System (UMTS), the 3G successor to GSM (Global System for Mobile communication). WCDMA promises high data speeds to mobile and portable wireless devices.

WCDMA is one of the technologies emerging from the FCS screening process as a promising FRS technology candidate. Previous detailed studies of this technology have investigated the performance of WCDMA to COCR service requirements and compatibility of this technology with existing aeronautical systems in L-band. This study of the COCR service provision in WCDMA provides another perspective of how to apply this technology to aeronautical systems.

3.1   Objective

The objective of this WCDMA study is to perform a functional analysis of the technology and identify how it can be used to support COCR services. Results of this analysis are used to determine the necessary elements of the architecture and protocol stack required to provision COCR services, which further supports the assessment of cost, certification, and standardization impact for applying WCDMA in aeronautical applications.

3.2   Approach

The overall approach applied for the WCDMA study consists of four steps. First, examine WCDMA network architecture, protocols and functions based on UMTS WCDMA network architectures and 3GPP technical specifications (ref. 10). Second, select COCR air traffic data link services for consideration based on the COCR and Radio Technical Commission for Aeronautics (RTCA)/DO–290 (ref. 11) documents. Third, map selected data link services to WCDMA network functions. Finally, results are documented in step 4. The flow of these tasks is illustrated in figure 24.

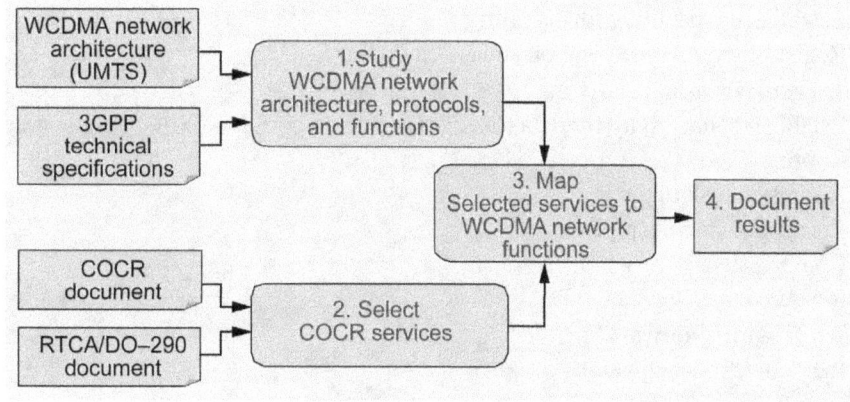

Figure 24.—Analysis approach.

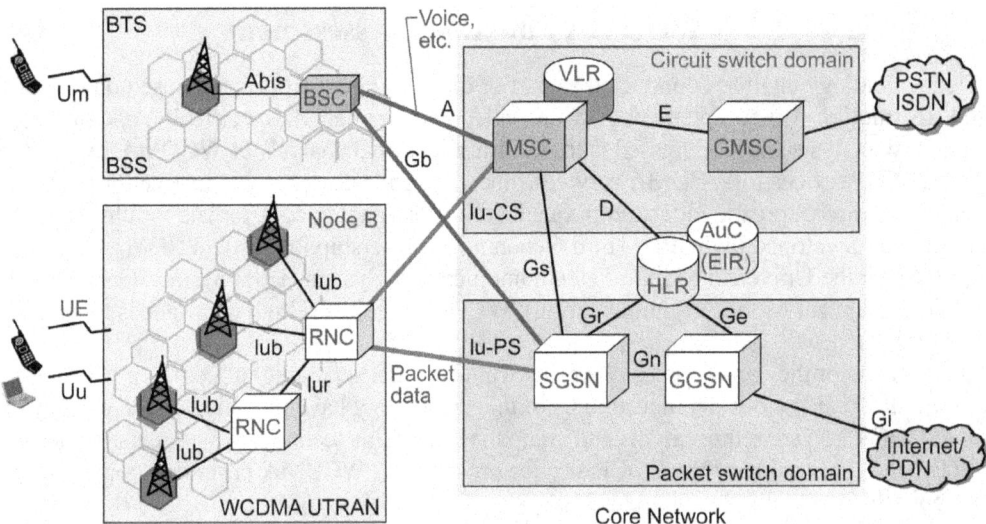

Figure 25.—WCDMA network architecture.

3.3 WCDMA Network Architecture

As shown in figure 25, a WCDMA network consists of three interacting domains: Core Network (CN), WCDMA UMTS Terrestrial Radio Access Network (UTRAN) and user equipment (UE). The UTRAN provides the air interface access method for UE. The radio base station is referred to as Node-B, and control equipment for Node-Bs is called the Radio Network Controller (RNC). The main functions of the core network are to provide switching, routing, and transit for user traffic. The core network also includes the databases and network management functions. The core network is divided in circuit switched (CS) and packet switched (PS) domains. Some of the CS elements are mobile switching center (MSC), visitor location register (VLR), and gateway MSC. Packet switched elements are serving GPRS support node (SGSN) and gateway GPRS support node (GGSN). Some network elements, like equipment identity register (EIR), home location register (HLR), VLR, and authentication center (AuC) are shared by both domains. The PS domain can connect to the internet or packet data network (PDN).

The WCDMA basic network function hierarchy is listed as follows:

- **Access network**
  - Finding access point
  - Localize user equipment
  - Establish connectivity for signaling
  - Registration + Authentication and Authorization
- **Route/transfer data**
  - Establish connectivity for data transfer (packet data protocol (PDP) context activation)
  - Routing
  - Address translating
  - Encryption and decryption
  - Compression
  - Resource management
- **Detach from the network**
  - PDP context deactivation
- **Mobility management**
  - Location management

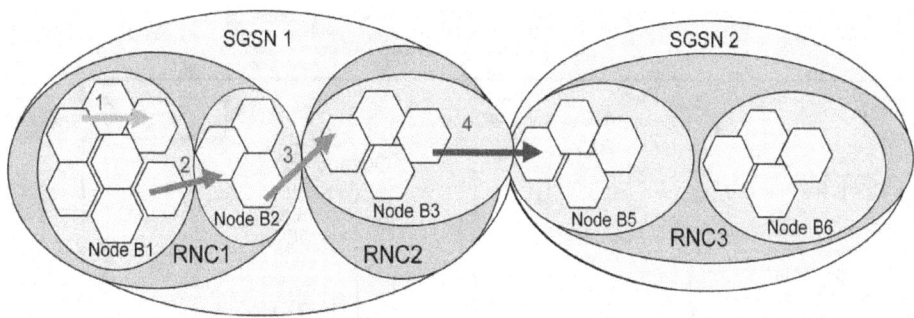

Figure 26.—Mobility management.

The PDP context activation function provides information to support packet delivery between a UE and the network. The PDP deactivation function results in deletion of PDP context information at RNC and SGSN.

Mobility management is the means by which a mobile network can keep track of the mobile subscriber location while connected to the network. The main function of mobility management is location management. In WCDMA, a group of cells is called a routing area (RA). The SGSN controls a service area containing several RAs. There are four location management procedures as shown in figure 26.

1. Cell Update—A UE informs the network of its current cell location.
2. Inter-Node B/Intra RNC/Intra-SGSN—A UE changes to a cell of different Node B but remains in the same RNC and SGSN service area.
3. Inter-SGSN Routing Update—Entry of a UE to a new RA triggers a change of SGSN service area.
4. Inter-RNC/Intra-SGSN Routing Update—A UE changes RA and triggers a change of RNC but remains serviced by the same SGSN.

3.4 Selection of Air Traffic Data Link Services

A set of air traffic data link services is selected for the analysis. These services support the implementation of aeronautical communications and are used to identify required WCDMA protocol and functional elements.

3.4.1 Sources of Service Information

Two sources of service information were used in the analysis: COCR as the primary source and RTCA/DO–290 as a secondary source of service descriptions. The COCR includes a complete set of FRS data link service definitions including service descriptions and operational context. RTCA/DO–290 is an older document that was defined to provide the minimum requirements for air traffic data link services in continental airspace. Some of the services described in RTCA/DO–290 were the basis for service definitions in the COCR. RTCA/DO–290 provides service transaction details beyond what is provided in the COCR such as operational methods, sequence diagrams, abnormal events, safety, performance, etc.

Figure 27 shows the COCR services organized into three major categories: ATS, AOC, and Network Management. For this study, the selection of data link services is focused on the ATS category, particularly on services that are also defined in the RTCA/DO–290 with more detailed descriptions.[2] This selection of ATS services correlating with services defined in RTCA/DO–290 is shown in red text in figure 27. These services include data link initiation capability (DLIC), air traffic control clearance message (ACM), air traffic control clearance (ACL), ATC microphone check (AMC), departure clearance (DCL), downstream clearance (DSC), digital-automatic terminal information service (D–ATIS), and flight plan consistency (FLIPCY). Full names of these services along with service dependencies are shown in figure 28. Here the DLIC service can be seen to be a prerequisite for all other services.

---

[2]Although the envisioned details for the FRS services defined in the COCR with the same name as the DO–290 services may be different, the service descriptions in DO–290 was still considered generally applicable and was used for this assessment.

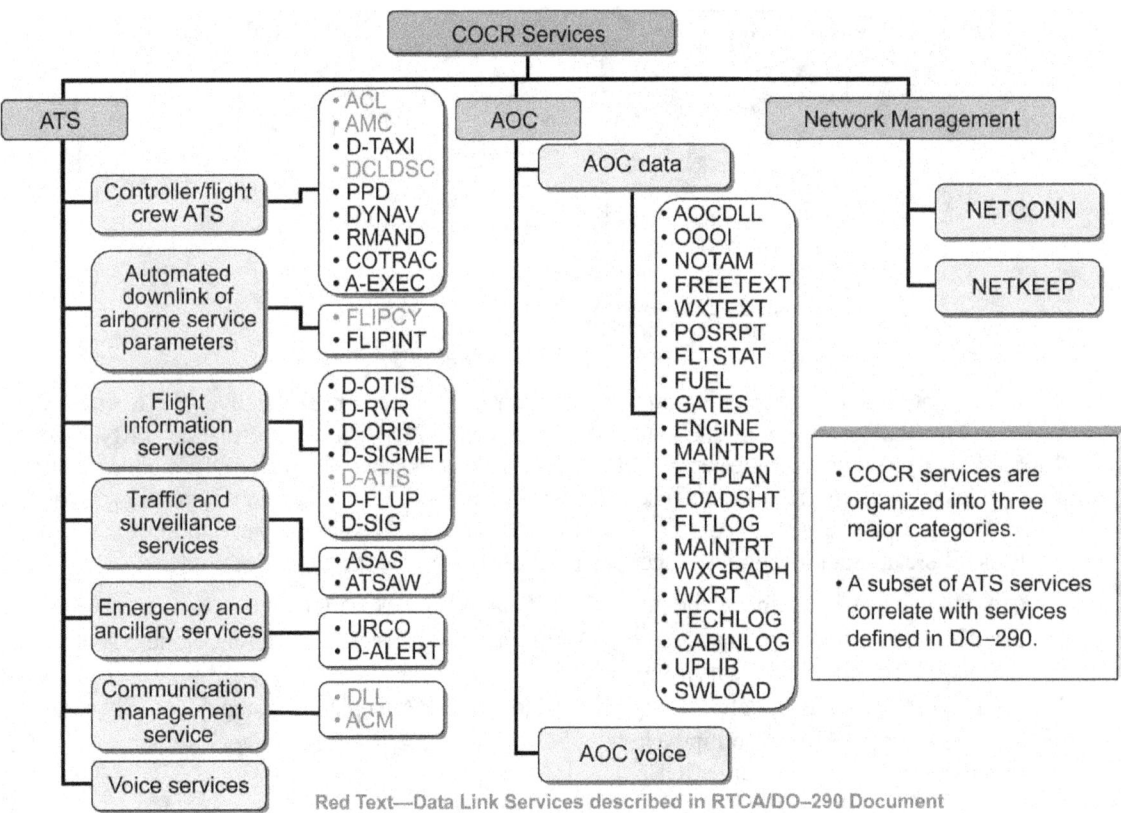

Figure 27.—COCR services.

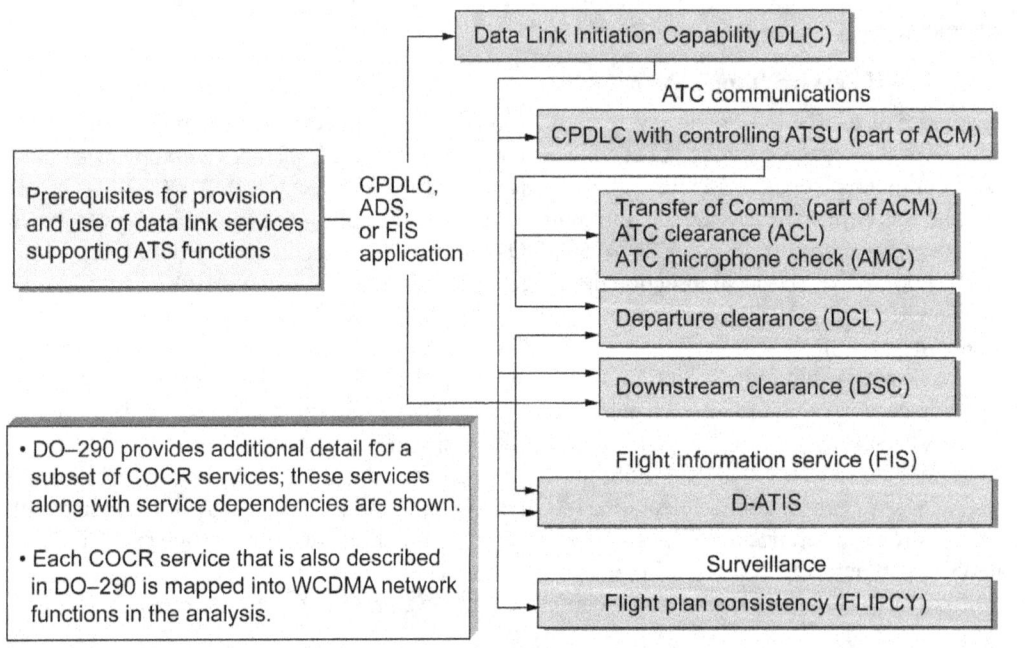

Figure 28.—Data link services in RTCA/DO–290.

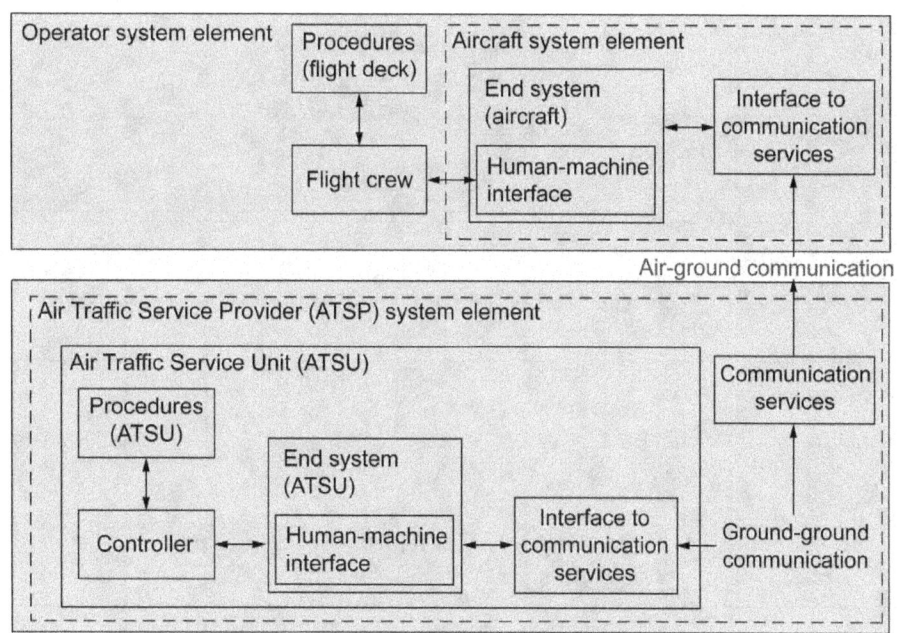

Figure 29.—CNS/ATM context diagram.

3.4.2 Mapping Services Into WCDMA Network Functions

The identified COCR services for consideration in this study are intended to be part of a future CNS/ATM implementation. Figure 29 (adopted from RTCA/DO–290) provides an overview of the CNS/ATM elements. These elements include the aircraft system, the air traffic service provider (ATSP) end systems, and the data link elements. Data link service elements may contain third-party communication services.

3.4.3 Logical Mapping

To provide a user's view of both the CNS/ATM elements and WCDMA network elements, a logical mapping of the two is provided as shown in figure 30. Here WCDMA network elements UE; UTRAN/RNC and Core network (PS domain); and PDN are mapped to Aircraft System Element; Interface to ATSP Communication Services/ATSP Communication Services; and Air Traffic Services Unit (ATSU), respectively. An air traffic context is also shown to reflect the potential application of WCDMA in CNS/ATM context.

3.5 Service Mapping

High level WCDMA network functions and DLIC service and its functions are described in further detail in this section. As noted previously, a set of eight COCR services was selected for the analysis. Although all eight services and functions are mapped into WCDMA interactions, as a representative example, only the mapping of DLIC service in WCDMA context is presented and explained in detail. Mapping results of the remainder of the data link services are presented only in terms of the service operational context and interaction sequences (app. B).

3.5.1 Logical Mapping

To provide a user's view of both the CNS/ATM elements and WCDMA network elements, a logical mapping of the two is provided as shown in figure 30. Here WCDMA network elements UE; UTRAN/RNC and Core network (PS domain); and PDN are mapped to Aircraft System Element; Interface to ATSP Communication Services/ATSP Communication Services; and

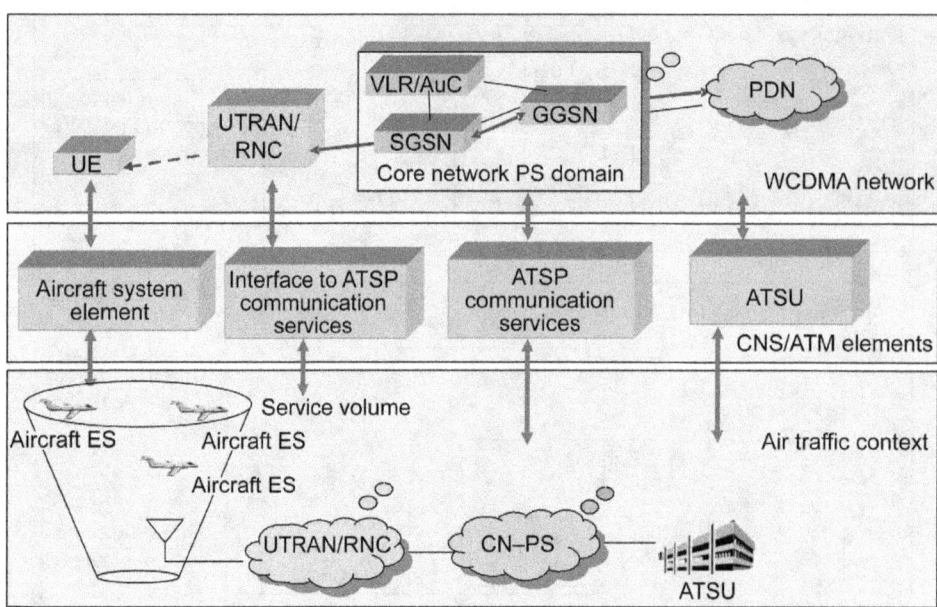

Figure 30.—Logical mapping.

ATSU, respectively. An air traffic context is also shown to reflect the potential application of WCDMA in CNS/ATM context.

3.6   Service Mapping

High level WCDMA network functions and DLIC service and its functions are described in further detail in this section. As noted previously, a set of eight COCR services was selected for the analysis. Although all eight services and functions are mapped into WCDMA interactions, as a representative example, only the mapping of DLIC service in WCDMA context is presented and explained in detail. Mapping results of the remainder of the data link services are presented only in terms of the service operational context and interaction sequences (app. B).

3.6.1   DLIC Service

The DLIC service is used to uniquely identify aircraft. The DLIC service exchanges information between an aircraft and an ATSU to support other data link services. The DLIC provides version and address information for all data link services including itself and this service is executed prior to any other data link services. The three functions of DLIC are

Initiation.—Exchange information and provide flight data to an ATSU. This function is between an aircraft and an ATSU. Information for other ATSU (up to eight) maybe exchanged.

Contact.—An ATSU provides the DLIC address of a specific ATSU to an aircraft and requests the initiation function be performed between the aircraft and the specified ATSU.

Update.—An ATSU provides its updated initiation information to an aircraft. The ATSU may update the information for other ATSUs when the ATSU has such initiation information.

The three DLIC functions exchanged between a moving aircraft and multiple ATSUs are illustrated in figure 31 (from RTCA/DO–290).

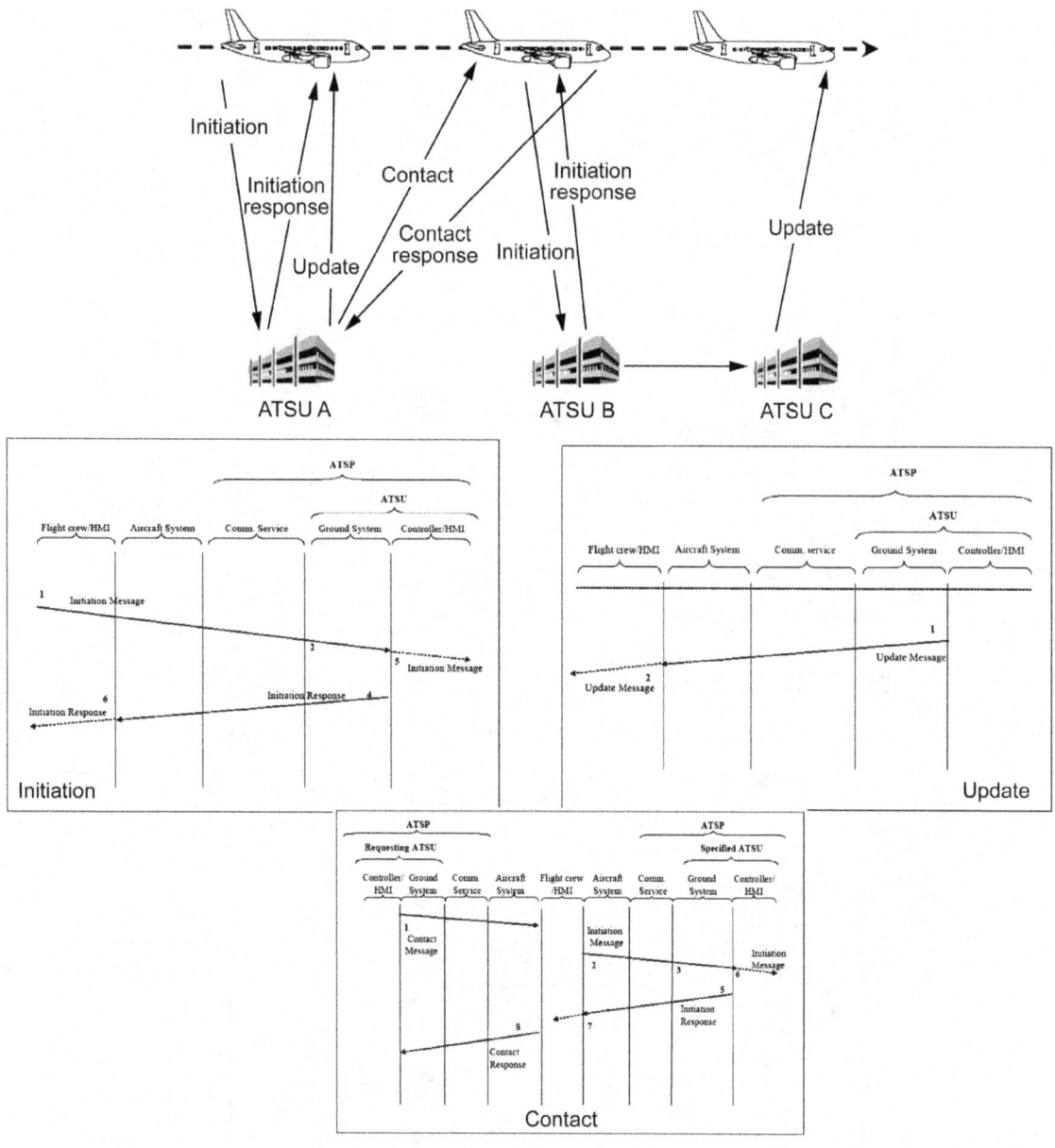

Figure 31.—DLIC operational context and interactions.

3.6.2 DLIC to WCDMA Services

Based on assessment of the WCDMA protocol, in order to provide DLIC capabilities, the following WCDMA network elements are needed: UE (aircraft end system), UTRAN/RNC, SGSN, HLR, GGSN, and ATSU End System. Sequence diagrams that illustrate the interactions among these elements are shown in figures 32 and 33, in which columns represent the necessary WCDMA elements and rows represent the three high level network functions exercised:

- Access Network function, which further includes the Find Access Point, Localize UE, Setup Radio Resource Control (RRC), and Registration with security functions
- Route/Transfer Data function (contains the PDP Context Activation and data transfer functions)
- Detach Network function (contains the PDP deactivation function)

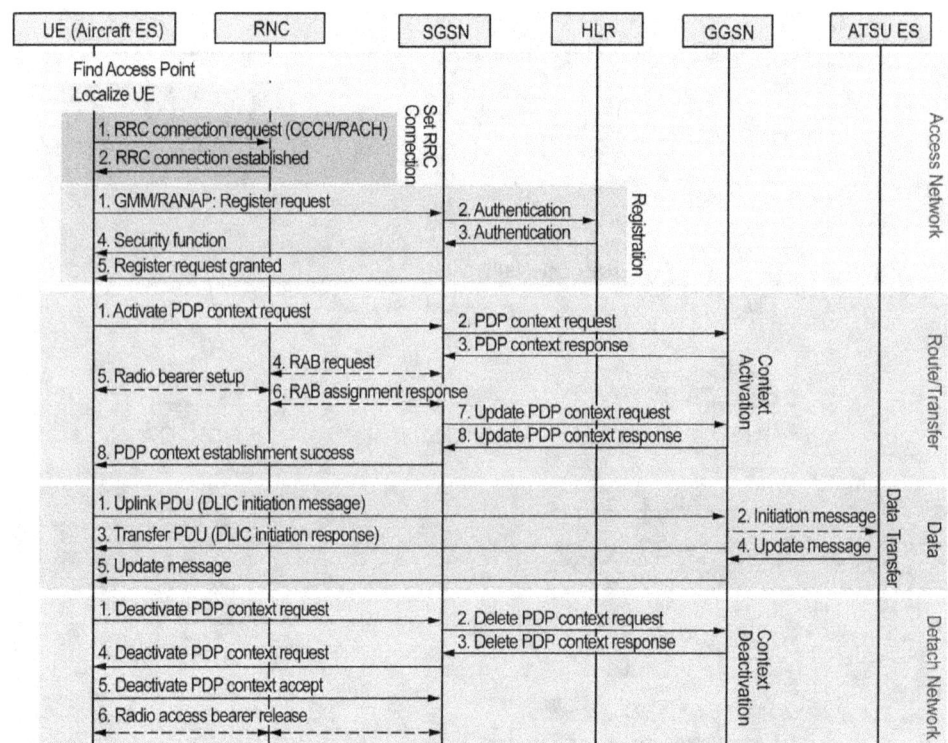

Figure 32.—Map DLIC Initation+Update functions to WCDMA service.

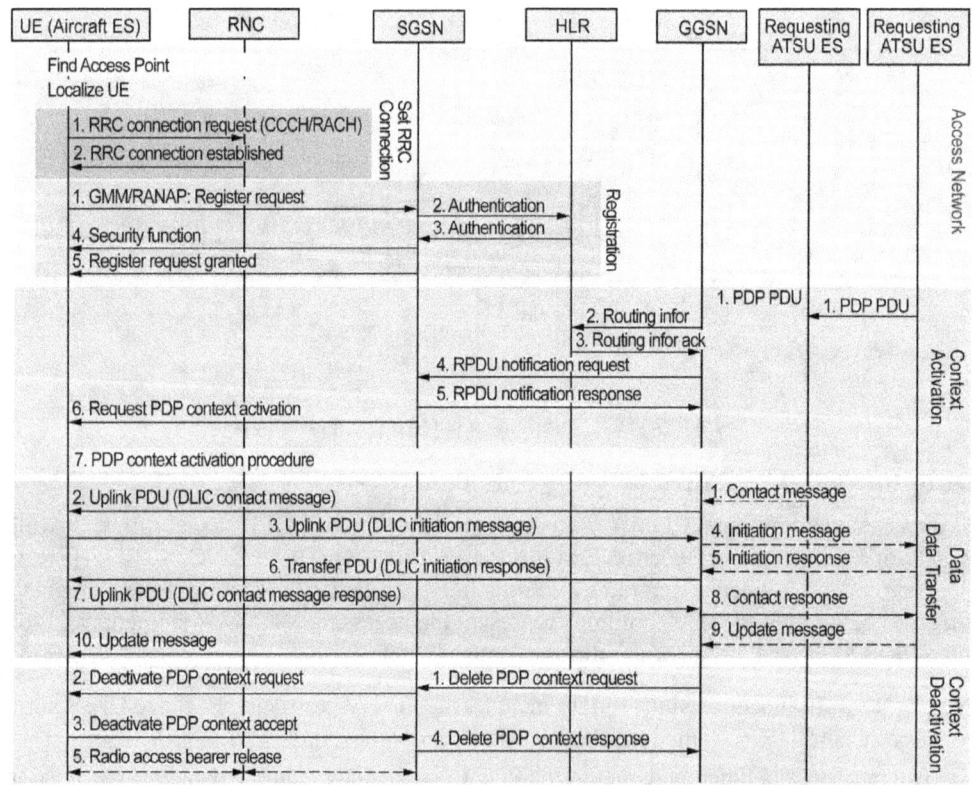

Figure 33.—Map DLIC Contact+Update functions to WCDMA service.

WCDMA specific services that support aircraft movement through the ATC system have also been examined. Four operational scenarios were identified and mapped to aircraft movement scenarios. Figure 34 shows these four operational scenarios. Scenario I is when an aircraft moves within the same Node B range; Scenario II is when the aircraft moves across different Node B but still within the same RNC service area; Scenario III is when the aircraft moves across different RNCs but within the same SGSN area; Scenario IV is when the aircraft moves across different SGSN and RNC service areas. WCDMA mobility management procedures such as the "Cell/ Universal Mobile Telecommunications System Terrestrial Radio Access Network Registration Area (URA) Update and RNC Relocation procedure" are needed to reflect these aircraft movement scenarios.

In Scenario IV, an aircraft moves across the SGSN area and needs to update from an old RNC and an old SGSN to a new RNC and a new SGSN service area. The network requires a mechanism to request location updates when a new SGSN service area is entered. Figure 35 shows a sequence diagram that reflects the Scenario IV cell update and RNC relocation mechanism.

3.6.3 Other Selected Data Link Services

The remainder of the data link service mappings can be found in appendix B.

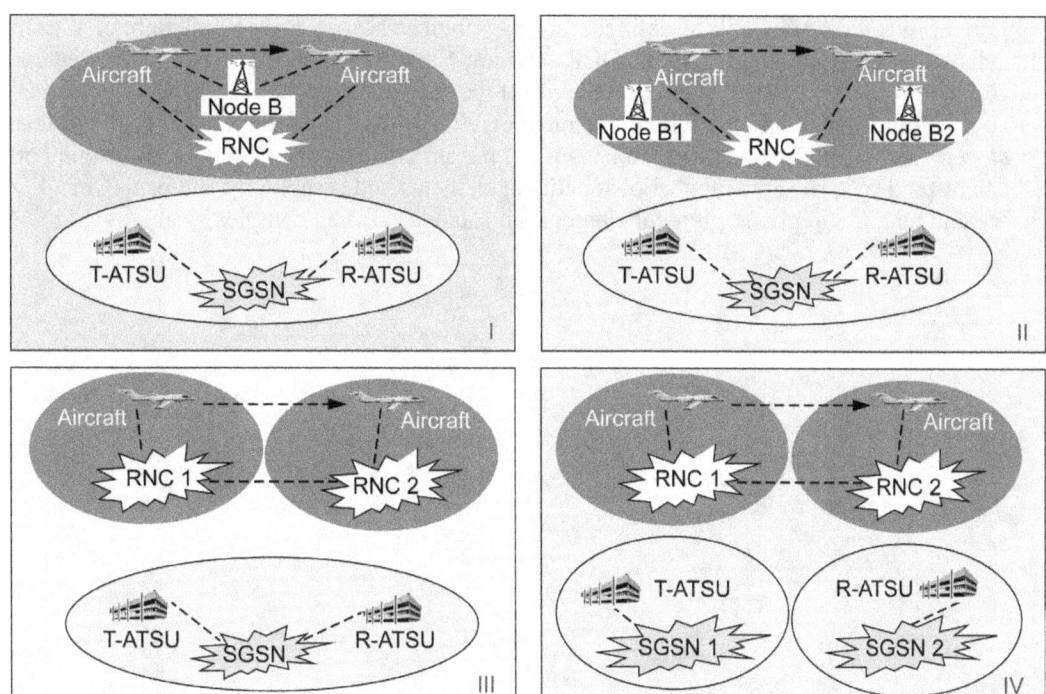

Figure 34.—Aircraft movement scenarios.

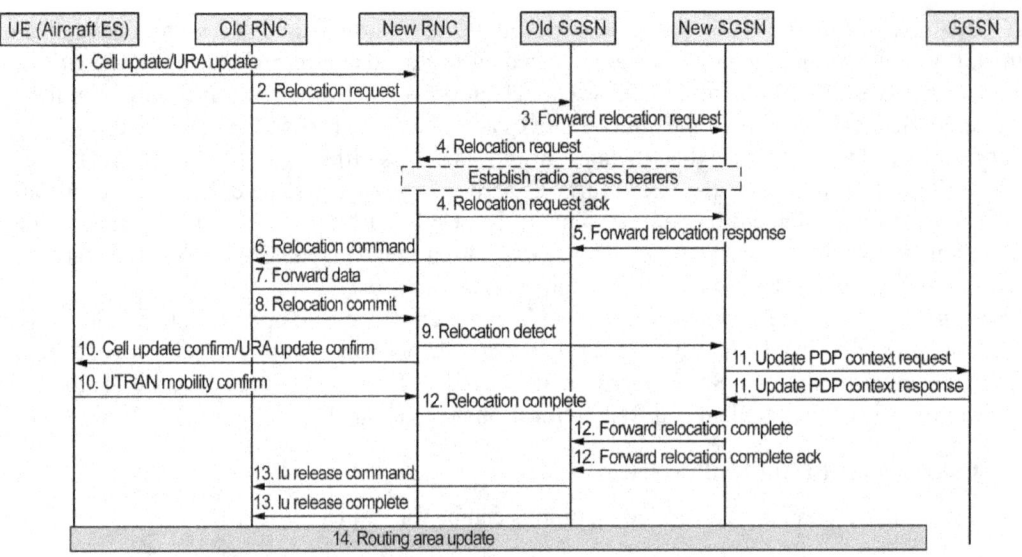

Figure 35.—Cell update and RNC relocation–Scenario IV.

3.7 Conclusions

Applying the WCDMA standards as defined, the study indicates that a full complement of WCDMA functional elements is required to provision COCR services. Not only the air interface and elements of the radio network controller are needed, but also elements of the core network such as HLR, SGSN, and GGSN. Required implementation of a full complement of WCDMA functionally elements and protocols affects cost, certification, and standardization because of the anticipated correlations between number of ground elements and cost, required number of functional elements and complexity and/or risk of certification, and number of ground/protocol elements and standardization complexity and/or risk.

4.0 P34/TIA–902 Intellectual Property Assessment

Phases I and II of the FCS recommended a set of technologies for further consideration for applicability to the FRS. This included P34, a set of public safety radio system standards. Because there are some patents associated with the P34/TIA–902 standards, it is informative to evaluate the potential impact of its intellectual property rights in the context of an FRS implementation. This topic is addressed in the following subsections.

4.1 Introduction to Intellectual Property

According to United States Patent and Trademark Office (USPTO)[3] (footnote 14):

> A U.S. patent for an invention is the grant of a property right to the inventor, issued by the United States Patent and Trademark Office. Generally, the term of a new patent is 20 years from the date on which the application for the patent was filed in the United States.
>
> The right conferred by the patent grant is, in the language of the statute and of the grant itself, "the right to exclude others from making, using, offering for sale, or selling" the invention in the United States or "importing" the invention into the United States. What is granted is not the right to make, use, offer for sale, sell or import, but the right to exclude others from making, using, offering for sale, selling or importing the invention. Once a patent is issued, the patentee must enforce the patent without aid of the USPTO.

4.2 Introduction to P34/TIA–902 Standards

APCO (Association of Public-Safety Communications Officers) Project 34 (P34) is an Electronics Industries Alliance/TIA standardized public safety radio system. The P34/TIA–02 concept was defined through a government and commercial partnership in order to provide universal access to all subscribers through a carefully controlled managed network. P34/TIA–902 addresses "issues that restrict the use of commercial services for mission critical public safety wireless applications" such as Priority Access and System Restoration, Reliability, Ubiquitous Coverage, and Security. P34/TIA–902 specifications are documented in the APCO Standards; they were developed by TR–8 Private Radio Technical Standards Committee under sponsorship of the TIA in accord with a memorandum of understanding between TIA and APCO/National Association of State Telecommunications Directors (NASTD)/FED.

P34/TIA–902 is considered a wideband public safety digital radio system that provides high-speed packet data services using the internet protocol (IP) on 50-, 100-, and 150-kHz channels and is designed for implementation in the 700 MHz band. The P34/TIA–902 system is defined with open, standardized interfaces so that manufacturers' equipment can be interoperable. It is the intent of the P34/TIA–902 specification that a system may be implemented where the equipment on either side of any open interface may be supplied by any manufacturer. The three open interfaces that are defined are the Wideband Air Interface (Uw); Data Peripheral Interface (Mobile Data Peripheral to Mobile Radio Control, Aw); and Data Interface (Radio Frequency Gateway to Data End System, Ew). The defined functional groups include the mobile radio, mobile routing and control, mobile data peripheral, base radio, base routing, and control and radio frequency gateway. P34/TIA–902 systems can provide connectivity between mobile radios and fixed network equipment (FNE); mobile radios to repeaters to mobile radios; and direct radio-to-radio (either mobile or fixed) connectivity. A depiction of the P34/TIA–902 open system architecture is shown in figure 36.

---

[3]Information is from USPTO Web site, http://www.uspto.gov/.

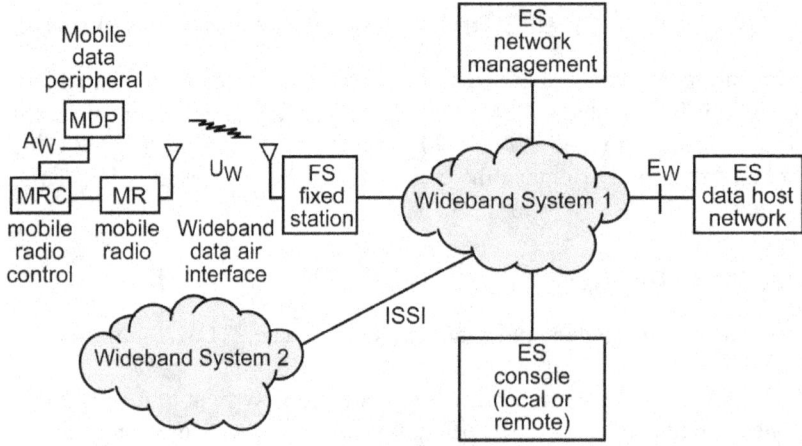

Figure 36.—P34/TIA–902 functional components and interfaces.

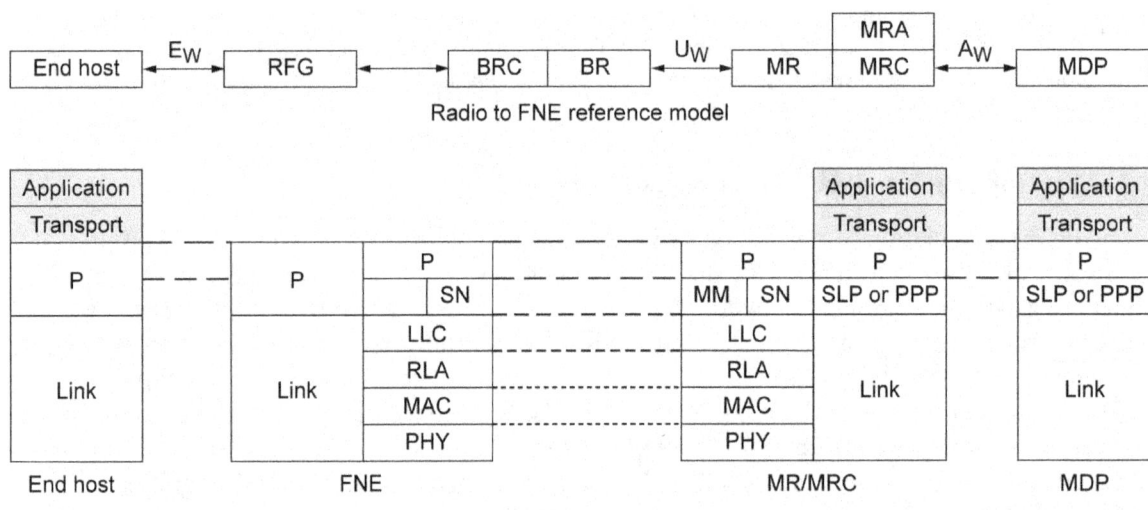

Figure 37.—P34/TIA–902 radio to fixed network equipment reference and protocol models.

The P34/TIA–902 protocol stack is layered, and assumes a point of attachment to an IP network. The P34/TIA–902 system is specified to provide IP version 4 (IPv4) and IPv6 bearer services for the transport of packet data using the IP suite of protocols. The wideband IPv4 (and IPv6) delivery service is required to directly support standard IP transport layers, including user datagram protocol (UDP), transmission control protocol (TCP) and real time protocol (RTP). This is shown in figure 37.

4.2.1   Patents in P34/TIA–902 Standards

There are eight patents associated with P34/TIA–902 standards, two of which are associated with Media Access Control/Radio Link Adaptation (MAC/RLA) layer specifications; six are in the physical (PHY) layer (Scalable Adaptive Modulation (SAM), channel coding (CHC), or IOTA). Table III lists these patents with patent name, patent number and date, patent assignee, associated P34/TIA–902 specification and protocol layer.

TABLE III.—PATENTS IN P34/TIA–902 STANDARDS

No.	Patent name	Patent number	Patent assignee	P34/TIA–902 specification	Protocol layer
1	Encryption Synchronization with Encryption Key Identification	US 5,185,796 Feb. 9, 1993	Motorola, Inc.	TIA–902.BAAC Wideband Air Interface MAC/RLA Layer Specification, Sept. 2002	MAC/RLA
2	Power Amplifier Linearization in a TDMA Mobile Radio System	US 5,559,807 Sept. 24, 1996	Motorola, Inc.	TIA–902.BAAC Wideband Air Interface MAC/RLA Layer Specification, Sept. 2002	MAC/RLA
3	Method for Providing and Selecting Amongst Multiple Data Rates in a Time Division Multiplexed System	US 5,533,004	Motorola, Inc.	TIA–902.BAAD–A Scalable Adaptive Modulation (SAM) Channel Coding Specification, Sept. 2003	SAM CHC
4	Communication Signal Having a Time Domain Pilot Component	US 5,519,730 May 21, 1996	Jasper	TIA–902.BAAB–A SAM Physical Layer Specification, Sept. 2003 TIA–902.BBAB	SAM & IOTA PHY
5	Peak to Average Power Ratio Reduction Methodology for QAM Communications Systems	US 5,381,449 Jan. 10, 1995	Motorola, Inc.	TIA–902.BAAB–A SAM Physical Layer Specification, Sept. 2003	SAM PHY
6	Quadrature Amplitude Modulation Synchronization Method	US 5,343,499 Aug. 30, 1994	Motorola, Inc.	TIA–902.BAAB–A SAM Physical Layer Specification, Sept. 2003	SAM PHY
7	Scalable Pattern Methodology for Multicarrier Communication Systems	US 6,424,678 Jul. 23, 2002	Motorola, Inc.	TIA–902.BAAB–A SAM Physical Layer Specification, Sept. 2003	SAM PHY
8	Construction of a Multicarrier Signal	US 6,278,686	France Telecom and Télédiffusion de France	TIA–902.BBAB Wideband Air Interface Isotropic Orthogonal Transform Algorithm Physical Layer Specification	IOTA PHY

4.3   Assessment Approach

As shown in figure 38, a six-step approach assessed P34/TIA–902 intellectual property. The first step is to identify patents listed in the P34/TIA–902 specifications; step 2 is to order these patents (textual technical descriptions) from the USPTO; in step 3, each of these patents were studied carefully and a high-level summary of each patent is presented for steps 4 and 5 reviews. In step 4, a corporate-level patent counsel was invited to review issues; meanwhile, impacts of patents to potential P34/TIA–902 application to FRS were addressed. In step 5, a team review was conducted to obtain inputs from different technical perspectives. Steps 4 and 5 were iterated until a consensus was achieved. Finally in step 6, recommendations for each patent were proposed and documented.

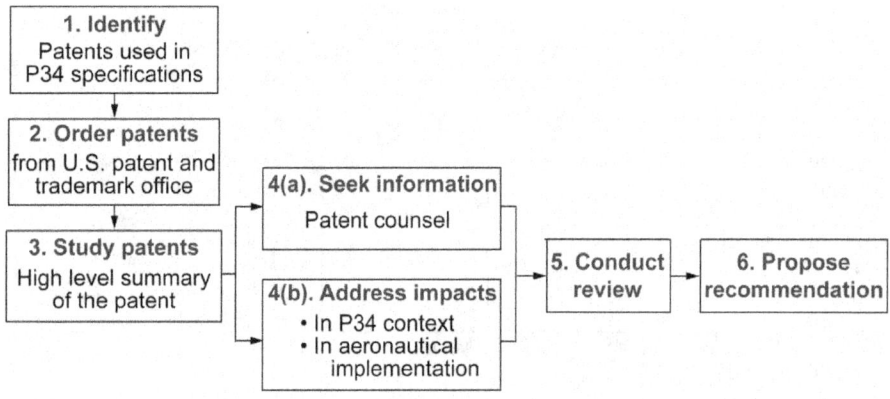

Figure 38.—Analysis approach.

4.3.1 Propose a Recommendation

For this study, recommendations specific to reviewed patents took the form of three implementation options, including

1. **Bypass the Patent:** This is the case when the patented features are not desirable applications and are considered to not be part of an FRS solution.

2. **Find an Alternative Solution:** In this case, the patented features are considered desirable, but could be implemented in a different manner; or only a subset of the functionality described in the patent is considered for implementation.

3. **Implement the Patent:** This is the case the identified patented feature is desirable, and there appears to be no alternative solution; thus, full implementation of the patent is expected.

Proposed recommendations were determined based on the desirability and criticality of the patents.

- **Desirability:** Are the claimed benefits desirable in FRS?
- **Criticality:** Can P34/TIA–902 be implemented without this patent? Will overall P34/TIA–902 performance be affected when the patent is not implemented? Is there an alternative solution?

4.3.2 Terms and Assumptions

In the process of proposing recommendations for the patents, certain terms and assumptions were found to be applicable. They are

1. All eight patents will all expire prior to FCS equipment deployment (assuming 2020 rollout). The term of a new patent is 20 years from the date on which the application for the patent was filed in the United States and is not renewable.

2. U.S. patents are not applicable to companies outside the United States. U.S. patents are not effective to implementations outside United States.

3. P34/TIA–902 physical layer modifications have been identified as needed for the application of this technology in the L-band aeronautical channel; six out of eight patents address features of the physical layer.

4. P34/TIA–902 has more flexibility than FRS applications may need; sometimes only a subset of the P34/TIA–902 characteristics is needed. Partial implementation of a patent is considered a "Alternative Solution" case (as the patent would not be applicable).

5. FRS ground infrastructure will likely be optimized for one modulation type, not for multiple modulation types, this reflects feedback received in FCS Phase I interim results in which the P34/TIA–902 concept of use included multiple modulation features was noted to adding unnecessary complexity and is not desirable.

6. Current analysis results suggest that quadrature phase skift-keying (QPSK) is likely sufficient to meet COCR requirements (quadrature amplitude modification (QAM) is not likely to be necessary).

4.4 Patent Assessments

Assessment results of the P34/TIA–902 associated patents are listed in this section. For each patent, the assessment result is given in the following form

- Name, number, assignee, and date
- Abstract of the patent
- Impact (assessment questions and answers regarding desirability and criticality of the patents)
- Proposed recommendation

Detailed descriptions of the patents are not available in this document, but can be requested from USPTO at http://www.uspto.gov/.

4.4.1 Patent 1

Name, number, assignee

Encryption Synchronization Combined with Encryption Key Identification
US 5,185,796, February 9, 1993, Motorola

Abstract of the patent

This patent describes a methodology to support the use of multiple encryption algorithms and/or encryption keys at MAC/RLA layer in a radio communication system.

Impact

Question	Answer
Are the claimed benefits desirable in FRS?	In COCR security requirements, cryptographic protection is identified as a preferred approach to mitigate three of the identified threats, and it states that this protection can be at the link layer, network layer, or application layer. This capability may be needed.
Can P34/TIA–902 be implemented without this patent?	Yes, when not use link layer encryption
Will overall P34/TIA–902 performance be affected when the patent is not implemented?	No, just drives higher layer requirements
Is there an alternative solution?	Possible, but not desirable for FRS

Recommendation: Bypass

4.4.2 Patent 2

Name, number, assignee

Power Amplifier Linearization in a TDMA Mobile Radio System
US 5,559,807, September 24, 1996, Motorola

Abstract of the patent

This patent describes a methodology for using predetermined linearization time slots in a TDMA radio system to transmit training signals so that receiver can linearize its power amplifier during that time slot. This approach avoids the use of substantial additional hardware and uses a minimum time allocated for training sequence and therefore causes minimum transmission delay.

Impact

Question	Answer
Are the claimed benefits desirable in FRS?	Yes
Can P34/TIA–902 be implemented without this patent?	No (requires framing structure change—not an insignificant change)
Will overall P34/TIA–902 performance be affected when the patent is not implemented?	Yes, functionally can be bypassed with re-work of MAC layer, but less capacity for data
Is there an alternative solution?	Probably yes, but it does require substantive changes to the frame structure in MAC layer

Recommendation: Implementation of the patent is desirable

4.4.3   Patent 3

Name, number, assignee

Method for Providing and Selecting Amongst Multiple Data Rates in a Time Division Multiplexed System
US 5,533,004, July 2, 1996, Motorola

Abstract of the patent

This patent defines an approach to TDMA that supports dynamic change in modulation to adjust to signal conditions. The approach combines multiple coding schemes with multiple modulation constellations that optimize TDMA time slot utilization resulting in maximum data throughput given channel conditions.

Impact

Question	Answer
Are the claimed benefits desirable in FRS?	Yes
Can P34/TIA–902 be implemented without this patent?	No (requires framing structure change—not an insignificant change)
Will overall P34/TIA–902 performance be affected when the patent is not implemented?	Yes, functionally can be bypassed with re-work of MAC layer, but less capacity for data
Is there an alternative solution?	Probably yes, but it does require substantive changes to the frame structure in MAC layer

Recommendation: Alternative Solution

4.4.4   Patent 4

Name, number, assignee

Communication Signal Having a Time Domain Pilot Component
US 5,519,730, May 21, 1996, Individual Inventors

Abstract of the patent

This patent describes a methodology for using time domain pilot symbols in a multicarrier communication system that allows for proper demodulation of information signals. Specific patent claims describe signal demodulation using pilot symbols that includes both phase and amplitude corrections.

Impact

Question	Answer
Are the claimed benefits desirable in FRS?	No, if use QPSK modulation
Can P34/TIA–902 be implemented without this patent?	Yes, if QPSK is the only modulation type used
Will overall P34/TIA–902 performance be affected when the patent is not implemented?	No, if QPSK is the only modulation type used
Is there an alternative solution?	Yes, other OFDM demodulation techniques should be evaluated if redesign P34/TIA–902 physical layer

Recommendation: Bypass

4.4.5   Patent 5

Name, number, assignee

Peak to Average Power Ratio Reduction Methodology for QAM Communication Systems
US 5,381,449, January 10, 1995, Motorola

Abstract of the patent

This patent describes a methodology to reduce peak to average power ratio by preselecting pilot amplitudes and phases for multiple subchannel, N-level QAM communication system

Impact

Question	Answer
Are the claimed benefits desirable in FRS?	No, if use QPSK modulation
Can P34/TIA–902 be implemented without this patent?	Yes, if QPSK is the only modulation type used
Will overall P34/TIA–902 performance be affected when the patent is not implemented?	No, if QPSK is the only modulation type used
Is there an alternative solution?	Possible, need further analysis, but not desirable for FRS

Recommendation: Bypass

4.4.6   Patent 6

Name, number, assignee

Quadrature Amplitude Modulation Synchronization Method
US 5,343,499, August 30, 1994, Motorola

Abstract of the patent

This patent describes a methodology for generating a synchronization sequence for a QAM communication system that reduces computational complexity required for data reception

Impact

Question	Answer
Are the claimed benefits desirable in FRS?	No, if use QPSK modulation
	(Yes, if use QAM and want to simplify receiver)
Can P34/TIA–902 be implemented without this patent?	Yes, if use QPSK modulation
Will overall P34/TIA–902 performance be affected when the patent is not implemented?	No for QPSK modulation
	Yes for QAM. Potential receiver cost may go up as receiver needs higher computational complexity
Is there an alternative solution?	Possible, need further investigation, but not desirable for FRS

Recommendation: Bypass

4.4.7 Patent 7

Name, number, assignee

Scalable Pattern Methodology for Multi-Carrier Communication Systems
US 6,424,678, July 23, 2002, Motorola

Abstract of the patent

This patent describes a methodology for scaling a base pattern of pilot symbols for various numbers of subchannels in a multicarrier communication system.

Impact

Question	Answer
Are the claimed benefits desirable in FRS?	No, no benefit in defining a scalable base pattern for a specific number of subchannels. FRS will pick one bandwidth instead if multiple subchannels and will optimize pilot pattern for that case
Can P34/TIA–902 be implemented without this patent?	Yes, a new symbol mapping figure will be required
Will overall P34/TIA–902 performance be affected when the patent is not implemented?	No, a new pattern can be defined and work as well
Is there an alternative solution?	Possible, need further investigation

Recommendation: Bypass

4.4.8 Patent 8

Name, number, assignee

Construction of a Multicarrier Signal
US 6,278,686, France Telecom & Télédiffusion De France, August 21, 2001

Abstract of the patent

This patent describes a methodology for creating a multicarrier communication signal by undergoing spectral processing to compress the signal in the time domain.

Impact

Question	Answer
Are the claimed benefits desirable in FRS?	Possible only if pulse transmission is desired
Can P34/TIA–902 be implemented without this patent?	No, but need to be validated with further research
Will overall P34/TIA–902 performance be affected when the patent is not implemented?	Yes, if use IOTA physical layer
Is there an alternative solution?	Unknown at this time

Recommendation: Bypass

4.5 Conclusions

From the analysis results shown in table IV, the following conclusions can be drawn:

- The concept of use defined for P34/TIA–902 makes some patents not applicable (for example, IOTA physical layer not used in the FCS application and associated patents do not apply); also recommended tailoring of physical layer standard for the FCS application results in bypassing of most physical layer patents.

- Only one patent is assessed as desirable to implement, it is a methodology proposed for power amplifier linearization, modification of which would influence MAC framing structure.

- Most if not all patents will expire before timeframe of FCS.
- These patents are not applicable to companies outside United States.
- Intellectual property associated with P34/TIA–902 standard is deemed to have little or no impact on the FRS if it is an implementation based on this standard.

TABLE IV.—PATENT EVALUATION SUMMARY

No.	Patent name	Patent number	Protocol layer	Decision	Comment
1	Encryption Synchronization Combined with Encryption Key Identification	US 5,185,796 Feb. 1993 Motorola Filed May 1991	MAC/RLA	Bypass	Implementation could be achieved by driving requirements to upper layers
2	Power Amplifier Linearization in a TDMA Mobile Radio System	US 5,559,807 Sept. 1996 Motorola Filed 1994	MAC/RLA	Implement patent desirable (expires 2014)	Alternative implementation could be achieved, but requires modification to MAC layer
3	Method for Providing and Selecting Amongst Multiple Data Rates in a Time Division Multiplexed System	US 5,533,004 Jul. 1996 Motorola Filed 1994	SAM Channel Coding	Alternative solution	Adaptive data rate not a desirable feature
4	Communication Signal Having a Time Domain Pilot Component	US 5,519,739 May 1996 Jasper Filed 1991	SAM & IOTA PHY	Bypass	Develop new PHY
5	Peak to Average Power Ratio Reduction Methodology for QAM Communications Systems	US 5,381,449 Jan. 1995 Motorola Filed 1991	SAM PHY	Bypass	Develop new PHY, and QAM is not identified modulation in CONUSE
6	Quadrature Amplitude Modulation Synchronization Method	US 5,343,499 Aug. 1994 Motorola	SAM PHY	Bypass	Develop new PHY, and QAM is not identified modulation in CONUSE
7	Scalable Patter Methodology for Multicarrier Communication Systems	US 6,424,678 Jul. 2002 Motorola Filed 2000	SAM PHY	Bypass	Develop new PHY
8	Construction of a Multicarrier Signal	US 6,278,686 Aug. 2001 France Telecom & Télédiffusion De France Filed 1996	IOTA PHY	Bypass	Develop new PHY

5.0 Summary of Results

This report documents the indepth studies that have been carried out during the third phase of the FCS technology investigation task. The studies included L-Band Interference Measurements; WCDMA Functional Assessment; and P34/TIA–902 Intellectual Property Assessment. A summary of results specific to each study are noted below.

L-Band Interference Measurements:

- The power levels expected from continuously transmitting FRS equipment onboard the aircraft may be sufficiently high as to cause desensitization in the DME interrogator. This phenomenon was evident for all of the FRS candidates even at large frequency separations for the DME that was tested. This finding is not favorable for FRS candidate technologies whose concept of use assumes continuous transmissions (e.g., WCDMA).

- The data also indicates that the DME interrogator is more tolerant to gated transmissions (i.e., there is potential for implementation of a technology with a gated waveform; but off-set channels may still be required (to be investigated)). A majority of the measurements used 100 percent duty cycles, which results in a conservative analysis. Lesser duty cycles may be expected in practice. It is expected that low-to-moderate duty cycles will interfere less with DME compared to FRSs with high duty cycles. This finding may be favorable for FRS candidate technologies whose concept of use assumes noncontinuous transmissions (i.e., LDL and P34/TIA–902 (partial)).

- It is recommended that further analysis be conducted to characterize the relationship between FRS duty-cycle and interference susceptibility (the duty-cycle investigation should include more variables than just overall duty cycle; there may be some combination of specific timescales of on/off pulses and overall duty cycle that results a seemingly "invisible" waveform from the DME interrogator's perspective); in the context of this investigation, identification of collocation constraints can also be investigated. It is also recommended that different models of DME interrogators be tested to provide a range of performance.

WCDMA Functional Assessment:

- Applying the WCDMA standards as defined, the study indicates that a full complement of WCDMA functional elements is required to provision COCR services. Not only the air interface and elements of the radio network controller are needed, but also elements of the core network such as HLR, SGSN, and GGSN. Required implementation of a full complement of WCDMA functionally elements and protocols has impact on cost, certification, and standardization because of the anticipated correlations between number of ground elements and cost, required number of functional elements and complexity and/or risk of certification, and number of ground/protocol elements and standardization complexity and/or risk.

P34/TIA–902 Intellectual Property Assessment:

- The concept of use defined for P34/TIA–902 makes some patents not applicable (e.g., IOTA physical layer not used in the FCS application and associated patents do not apply); also recommended tailoring of physical layer standard for the FCS application results in bypassing of most physical layer patents.

- Only one patent is assessed as desirable to implement, it is a methodology proposed for power amplifier linearization, modification of which would influence MAC framing structure.

- Most if not all patents will expire before timeframe of FCS.

- These patents are not applicable to companies outside the United States.
- Intellectual property associated with P34/TIA–902 standard is deemed to have little or no impact on the FRS if it is an implementation based on this standard.

The results of the studies noted above will be used to support the final technology evaluation and technology recommendation development effort, the last component of the FCS technology investigation study. The final FCS technology investigation report, to be published in early 2008, will document the overall technology evaluation results and subsequent recommendations.

Appendix A—Unfiltered Interference Measurement Results

The measurements results continued in this appendix are the unfiltered measurement results that were collected during the first pass of the measurement campaign. The filtered measurements can be found in Section 2. Figures 39, 40, and 41 show the interference measurements.

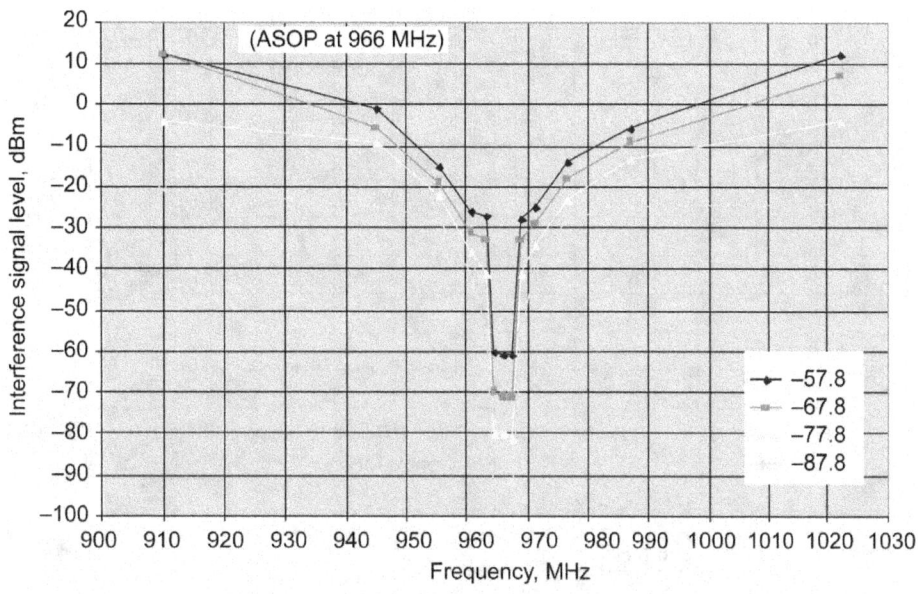

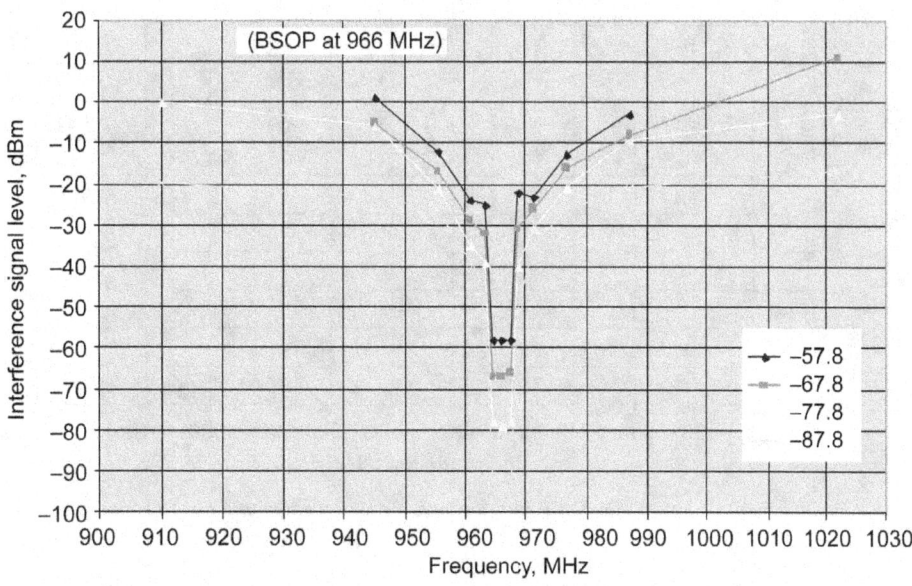

Figure 39.—WCDMA/DME interference measurements without filtering.

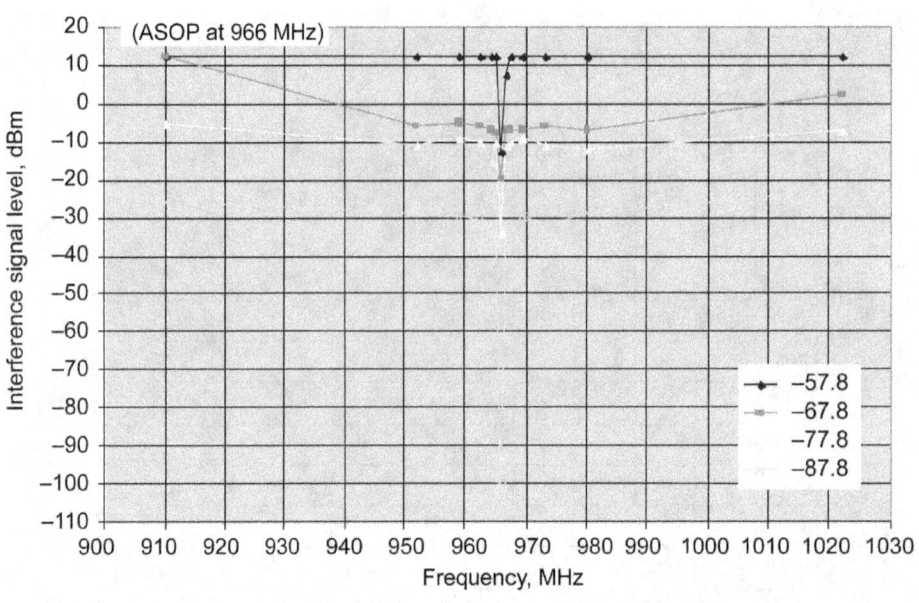

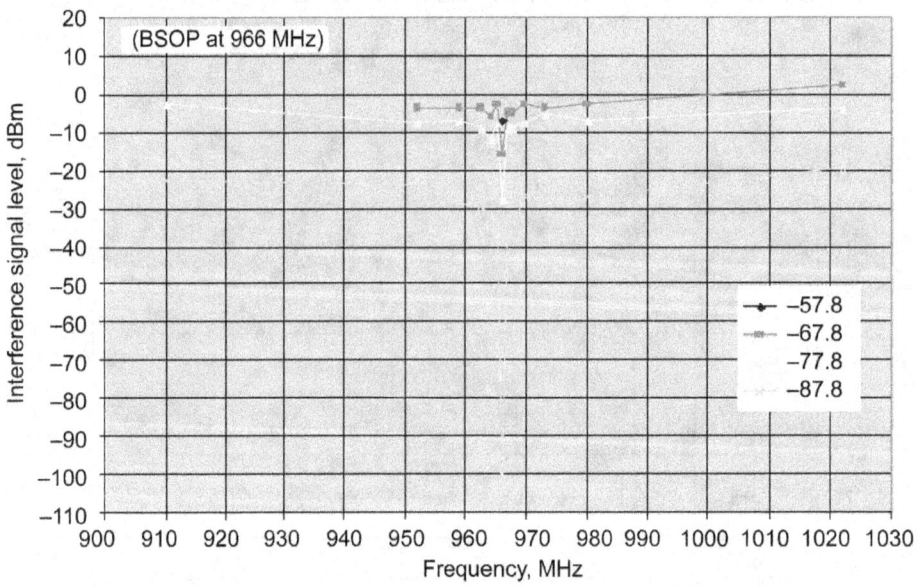

Figure 40.—Gated LDL/DME interference measurements without filtering.

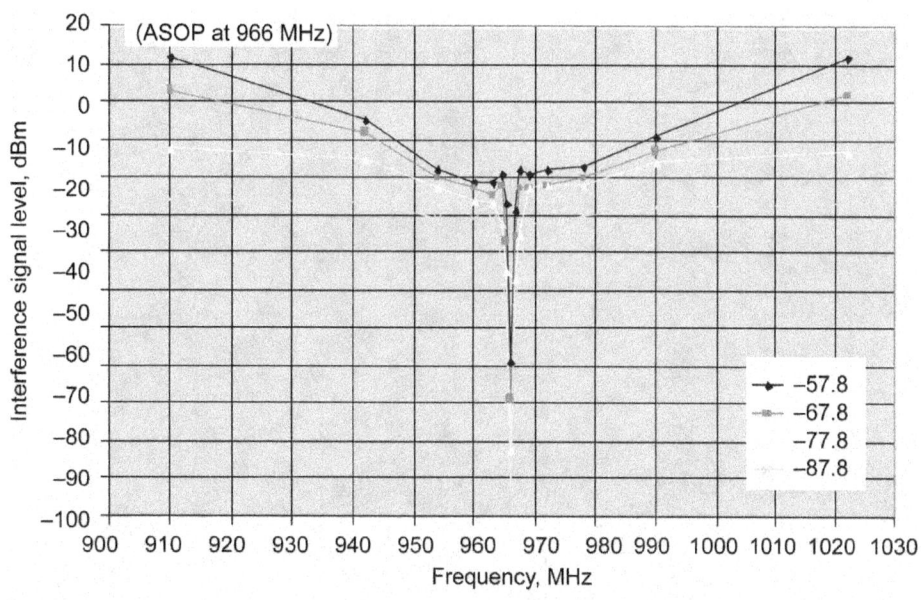

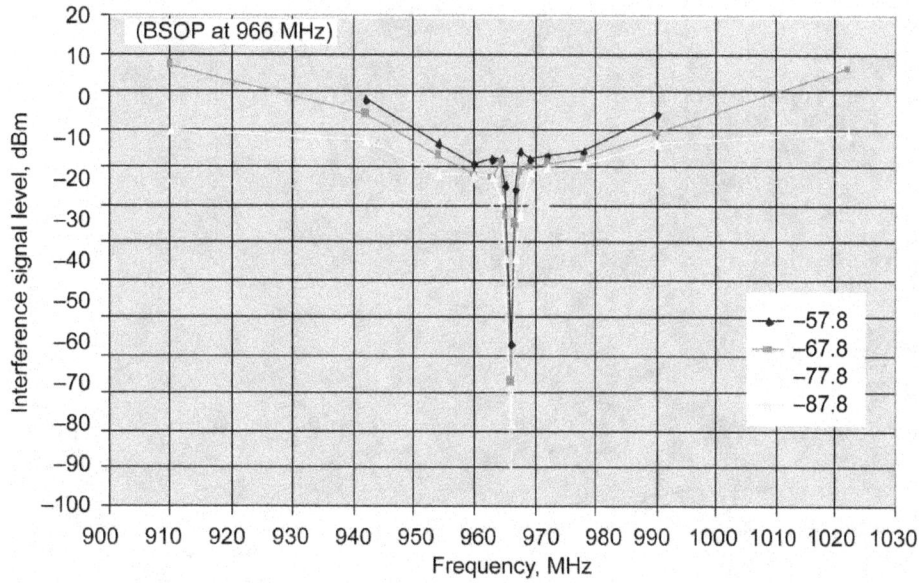

Figure 41.—P34/TIA–920 (50 kHz)/DME interference measurements without filtering.

Appendix B—Additional WCDMA Functional Analysis Results

This appendix provides additional WCDMA functional analysis result as described in Section 3.

B.1   ACM (ATC Communication) Service

CPDLC (controller pilot data link communication) is a means of communication between controller and pilot, using data link for ATC communications.

- CPDLC allows two-way communications between pilot and controller, but using data link rather than voice. It is especially useful in areas where difficult-to-use high-frequency (HF) radiotelephony was previously used for communication with aircraft, for example. oceans.

The data link service for ATC is enabled through the use of CPDLC, ACM is used to

- Establish CPDLC with an ATSU
- Terminate CPDLC with an ATSU
- Transfer voice comm and CPDLC form the T-sector/T-ATSU to the R-sector/R-ATSU
- Issue a change of frequency when this happens:
  - Initiated by the T-sector/T-ATSU, or
  - A request from the R-sector/R-ATSU, or
  - A request from the flight crew

Possible operating methods for ACM service are

- Transfer from T-ATSU to R-ATSU, both using CPDLC
- Transfer from T-ATSU not using CPDLC to R-ATSU using CPDLC
- Transfer from T-ATSU using CPDLC to R-ATSU not using CPDLC
- Transfer and/or change of frequency with no change in the established CPDLC, both sectors using CPDLC
- Transfer from T-sector not using CPDLC to R-sector using CPDLC, with no change in the established CPDLC
- Transfer from T-sector using CPDLC to R-sector not using CPDLC, with no change in the established CPDLC
- Termination of CPLDC without transfer of CPDLC or change of frequency

Three ACM functions exchanged between a moving aircraft and multiple ATSUs are illustrated in figure 42.

Two ACM Service Mapping Scenarios
- Mapping Scenario 1
  - ACM service operating method (a)
    - o   Transfer from T-ATSU to R-ATSU, both using CPDLC
  - Network Domain (II)
    - o   Same RNC routing area and same RNC region
- Mapping Scenario 2
  - ACM service operating method (a)
    - o   Transfer from T-ATSU to R-ATSU, both using CPDLC
  - Network Domain (IV)
    - o   Different RNC routing area and different SGSN region

A routing area update should take place when a WCDMA UE (aircraft) detects that it has entered a new routing area.

Mapping of scenario 1 and scenario 2 to WCDMA functions are shown in figures 43, 44, and 45.

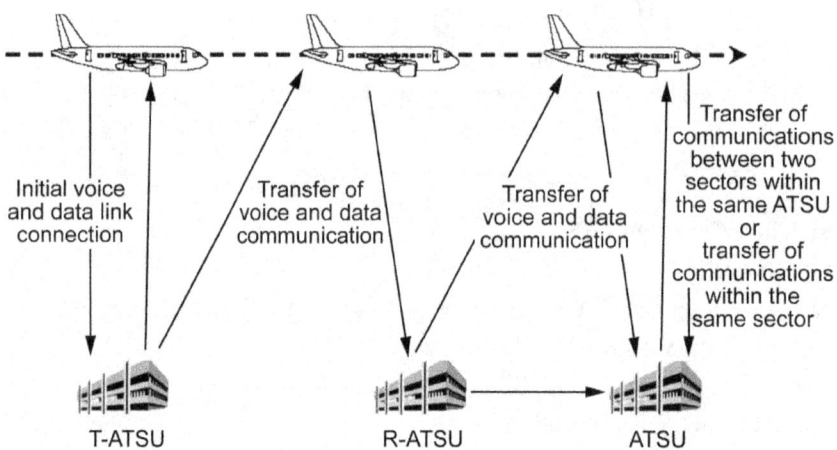

Figure 42.—ACM operational context.

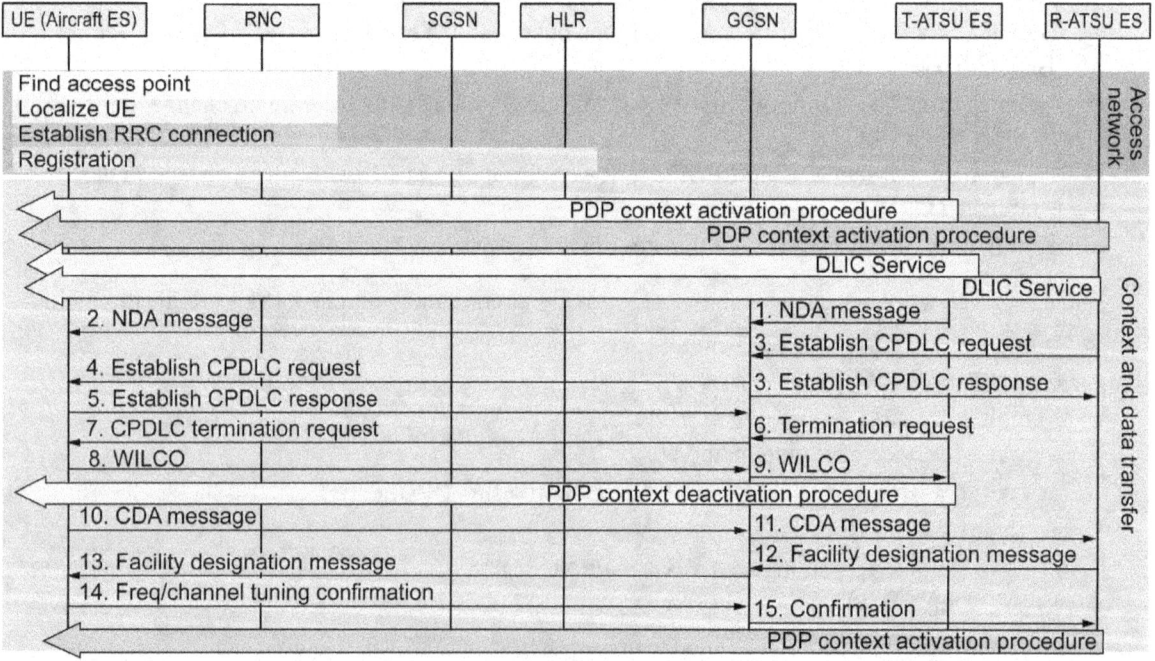

Figure 43.—ACM Scenario 1 in WCDMA service.

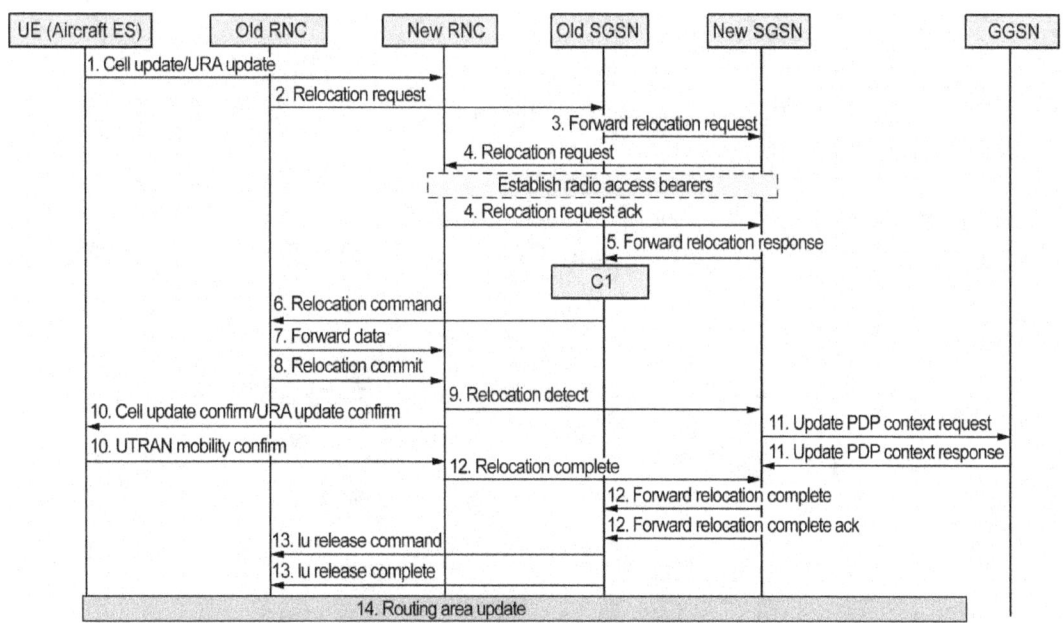

Figure 44.—ACM scenario 2 in cell/URA update and RNC relocation.

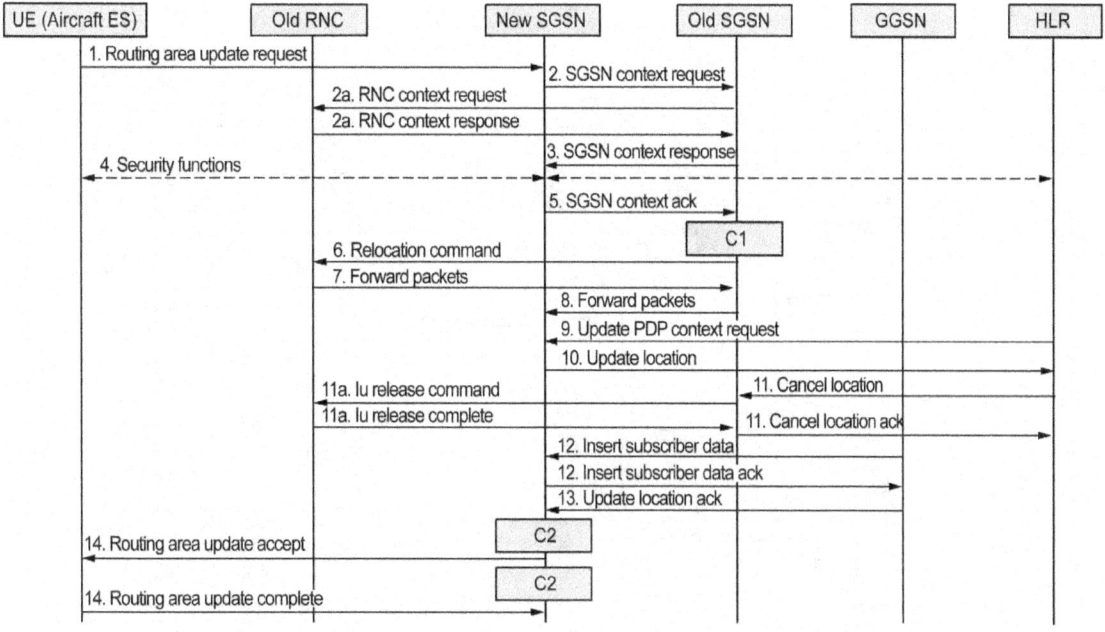

Figure 45.—ACM scenario 2 in inter-SGSN routing area update.

B.2   ATC Clearance (ACL) Service

ACL service supports operational communication between the flight crew and the controller of the Current Data Authority ATSU. ACL service is preceded by ACM Service, operating methods are

- Request clearances

- Issue and expect clearances
- Issue requests for current or future flight status
- Provide advisory information such as meteorological or operational conditions
- Provide flight status notifications

The operational context of ACL service is shown in figure 46. ACL functions exchanged between a moving aircraft ATSU are illustrated in figure 47. Mapping of ACL Clearance function into WCDMA service is shown in figure 48.

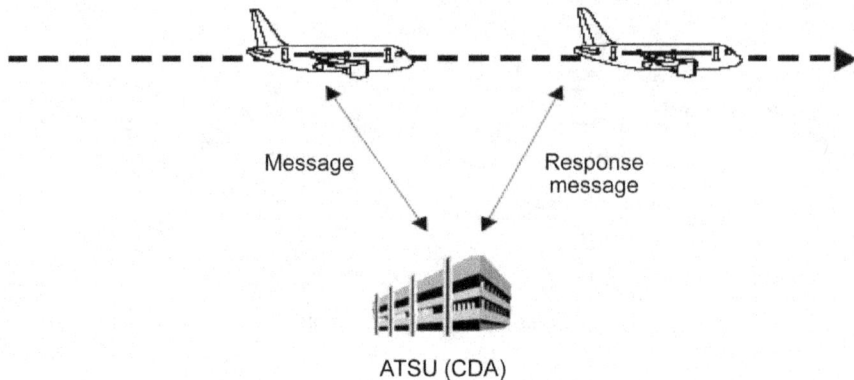

Figure 46.—ACL operational context.

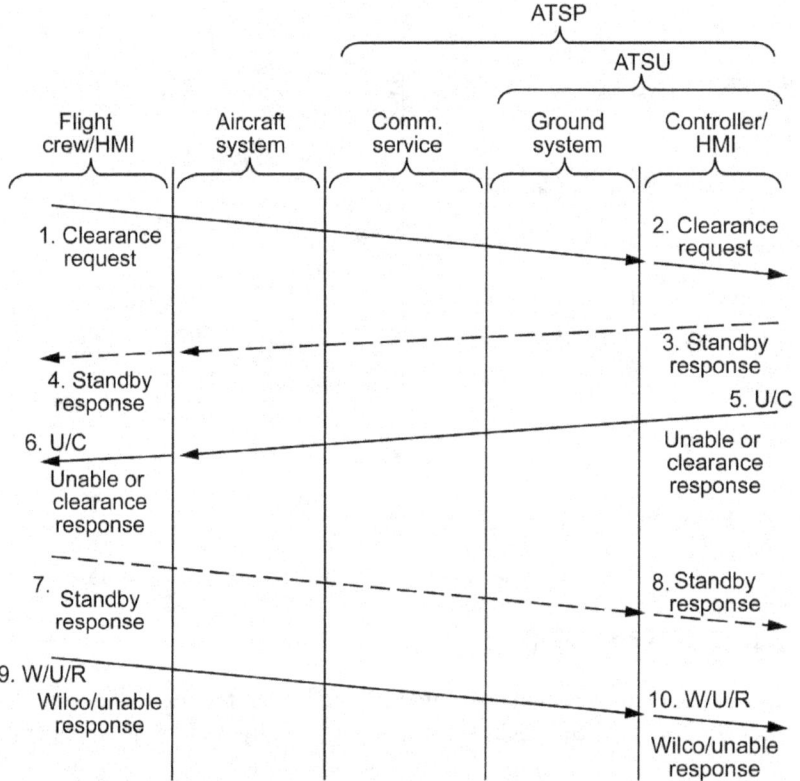

Figure 47.—ACL sequence diagram.

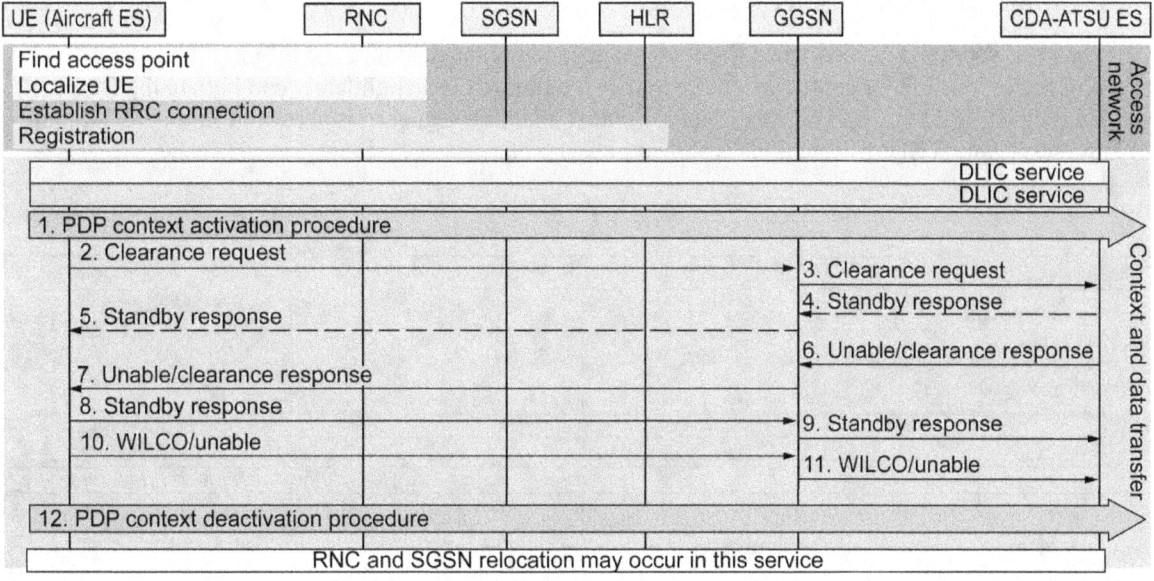

Figure 48.—ACL clearance in WCDMA service.

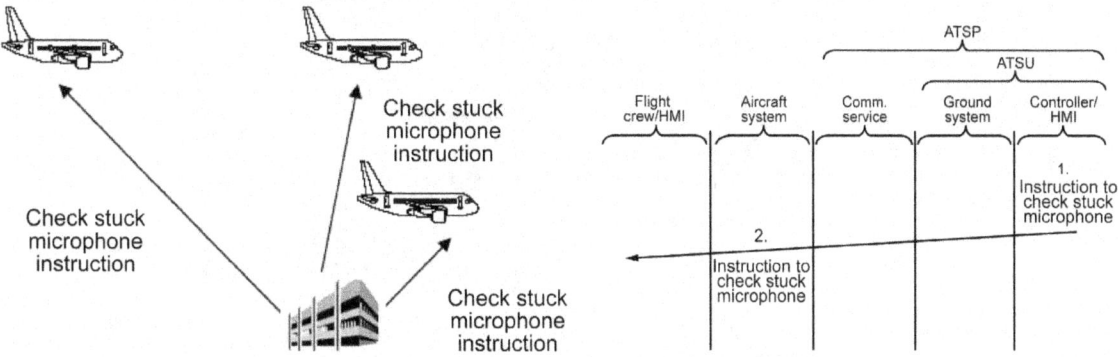

Figure 49.—AMC operational context and time sequence.

B.3 AMC (ATC Microphone Check) Service

The AMC service provides controllers with the capability to uplink an instruction to an aircraft in order for the flight crew to check that the aircraft is not blocking a given voice channel. The AMC operational context and time sequence is shown in figure 49. Mapping of AMC service to WCDMA functions is shown in figure 50.

B.4 Departure Clearance (DCL) Service

DCL provides automated assistance for requesting and delivering clearances, when DCL is provided through the use of CPDLC, this information can be exchanged using CPDLC message elements. DCL service is intended for use during the surface phase of operation. The DCL operational context and time sequence is shown in figure 51. Mapping of initial DCL service to WCDMA functions is shown in figure 52. DCL clearance revision in WCDMA context is shown in figure 53.

B.5   Downstream Clearance (DSC) Service

The DSC service is needed when flight crews need to obtain clearances or information from ATSUs that will be responsible for control of the aircraft in the future. Only flight crew can initiate the DSC service with a D-ATSU. The DSC operational context and time sequence is shown in figure 54. Mapping of DSC service to WCDMA functions is shown in figure 55.

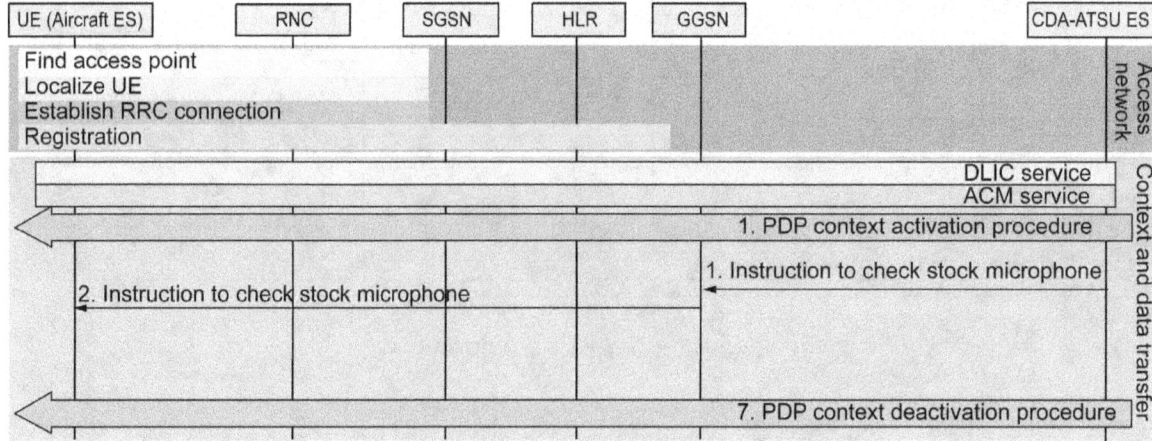

Figure 50.—AMC to WCDMA service.

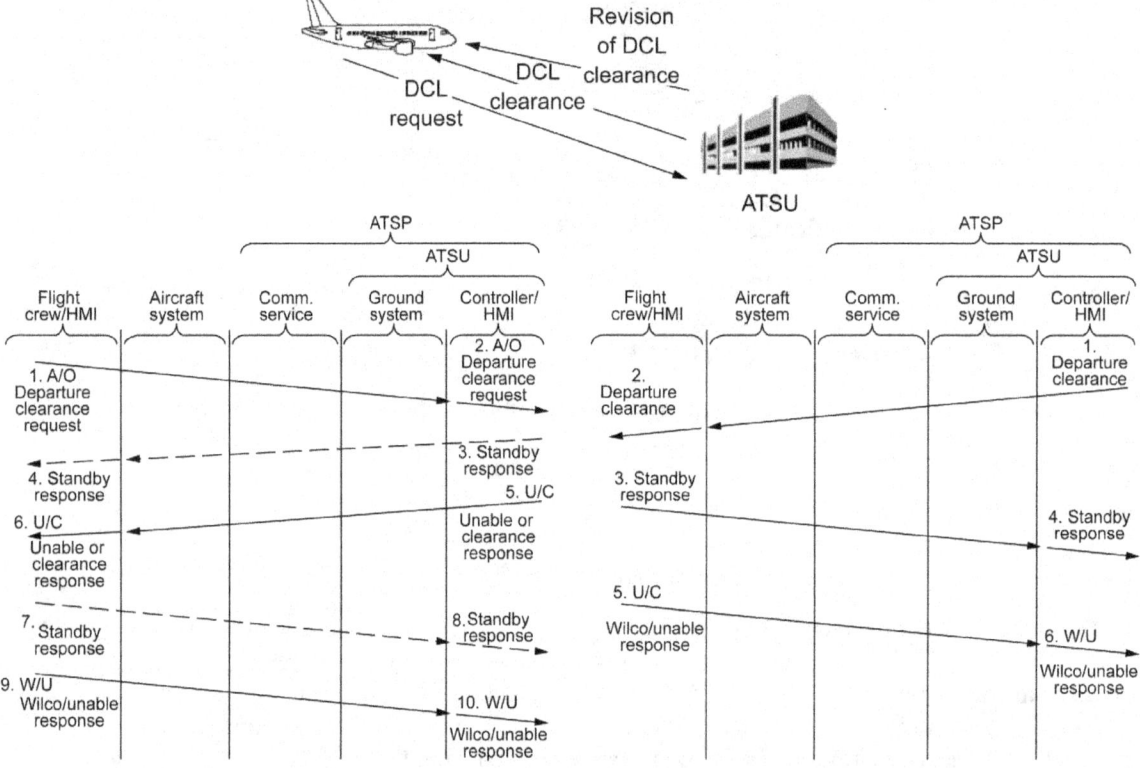

Figure 51.—DCL operational context and time sequence.

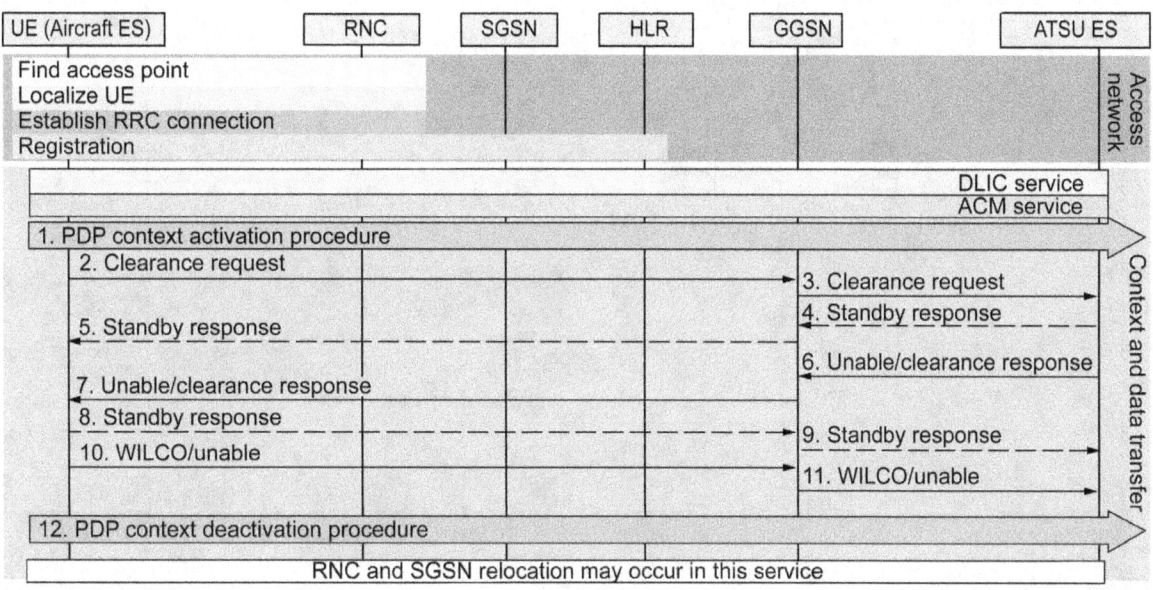

Figure 52.—Provision of initial DCL in WCDMA context.

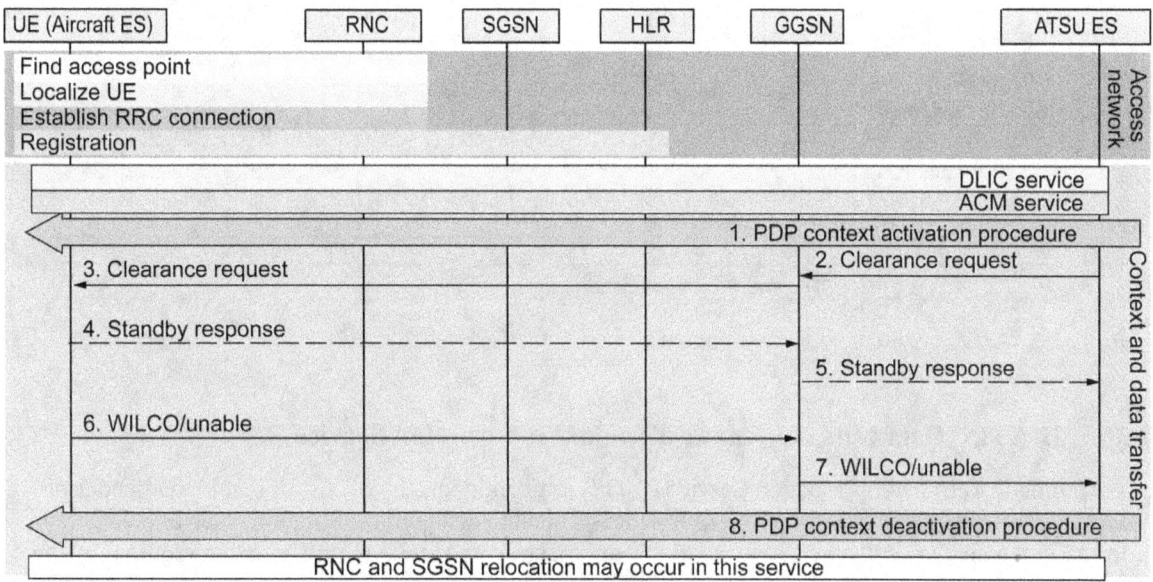

Figure 53.—DCL clearance revision in WCDMA.

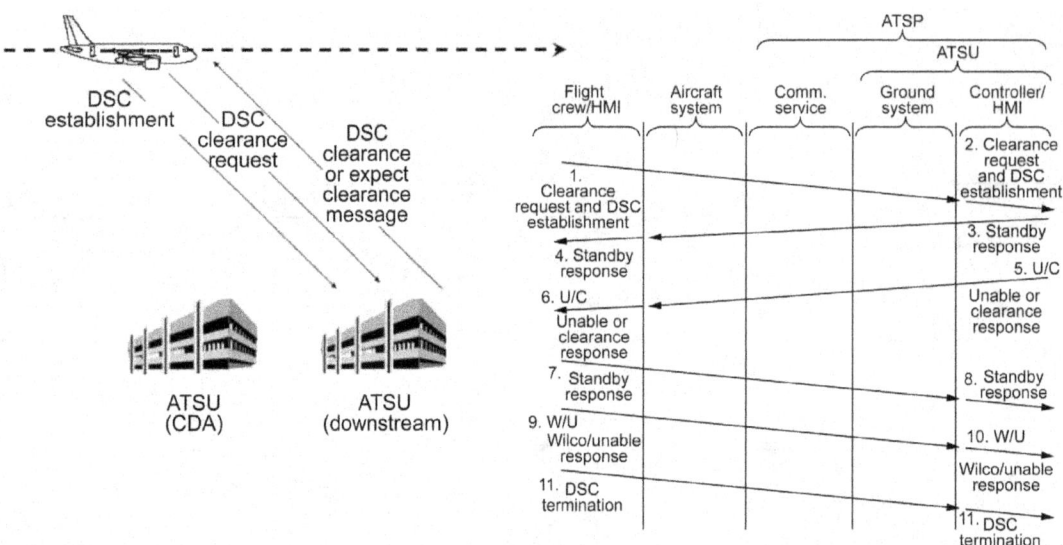

Figure 54.—DCL operational context and time sequence.

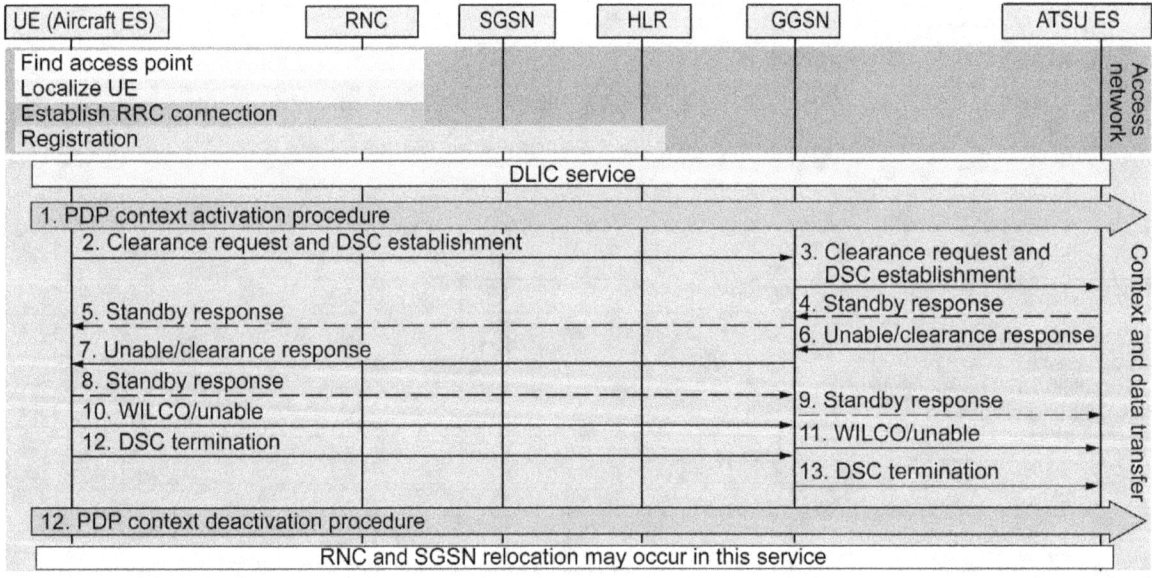

Figure 55.—DCL to WCDMA.

B.6   D-ATIS (Data Link Automatic Terminal Information Service)

Automatic Terminal Information Service (ATIS) is a continuous broadcast of recorded noncontrol information in busier terminal areas. ATIS broadcasts contain essential information, such as weather information, which runways are active, available approaches, and any other information required by the pilots, such as notices to airmen (NOTAMs). Pilots usually listen to an available ATIS broadcast before contacting the local control tower, in order to reduce the controllers' workload and relieve frequency congestion. The D-ATIS automatically generates ATIS information and also provides the ability to send the same information to the pilots using digital texts. This service is installed to reduce workload on the air traffic controllers as well as reduce the chances of misunderstandings.

Two modes:

- Demand mode—a single ATIS report requested, or
- Contract mode—an ATIS report followed by updates when they occur

The DCL operational context is shown in figure 56. D-ATIS demand and contract mode time sequence is shown in figure 57. D-ATIS function in WCDMA context is shown in figure 58.

B.7   FLIPCY (Flight Plan Consistency) Service

FLIPCY is used to automatically detect inconsistencies between the ATC flight plan and the flight plan activated in the aircraft system. It provides controllers with automated support confirming that the en route and terminal area portions of the aircraft system and ATSU flight plans conform. If an inconsistency is detected, the controller is informed.

The service may be initiated

- Before the aircraft takeoff and/or
- A parameter amount of time before entering the ATSU area of responsibility and/or
- At the time of entry into the ATSU's area of responsibility and/or
- At controller's discretion

FLIPCY requires that DLIC be performed to determine the aircraft's capability to accomplish the service.

The FLIPCY operational context and time sequence is shown in figures 59 and 60. Mapping of FLIPCY service to WCDMA functions is shown in figure 61.

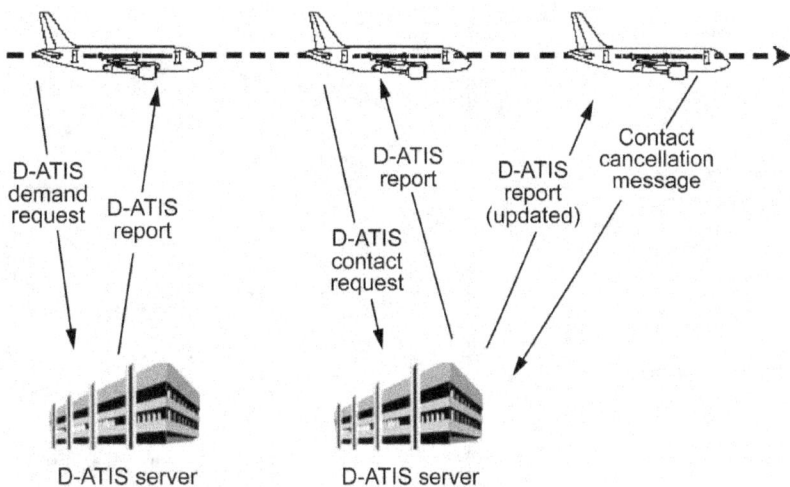

Figure 56.—D-ATIS operational context.

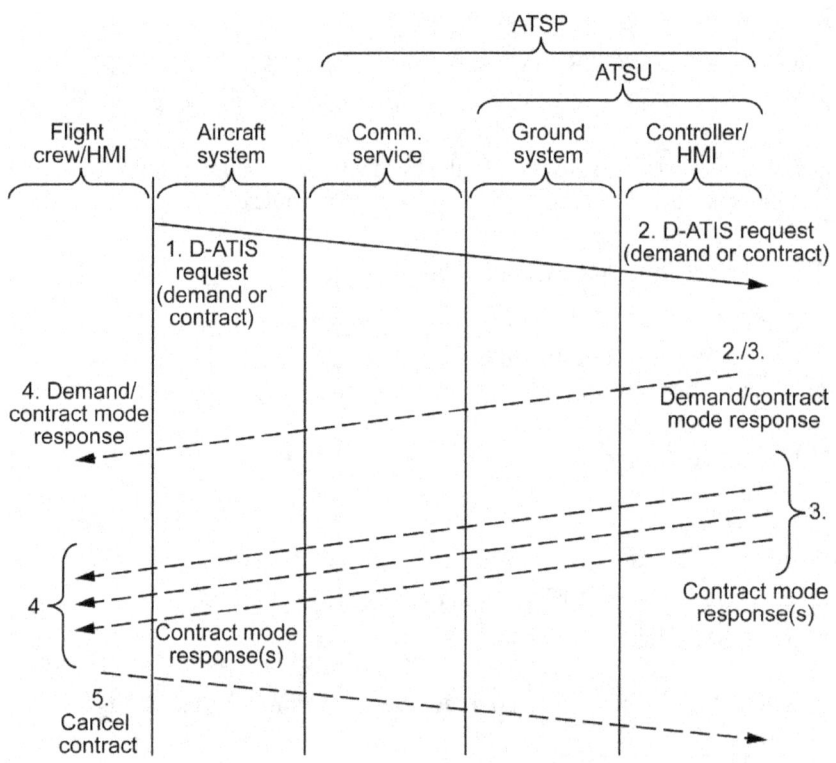

Figure 57.—D-ATIS demand and contract mode time sequence.

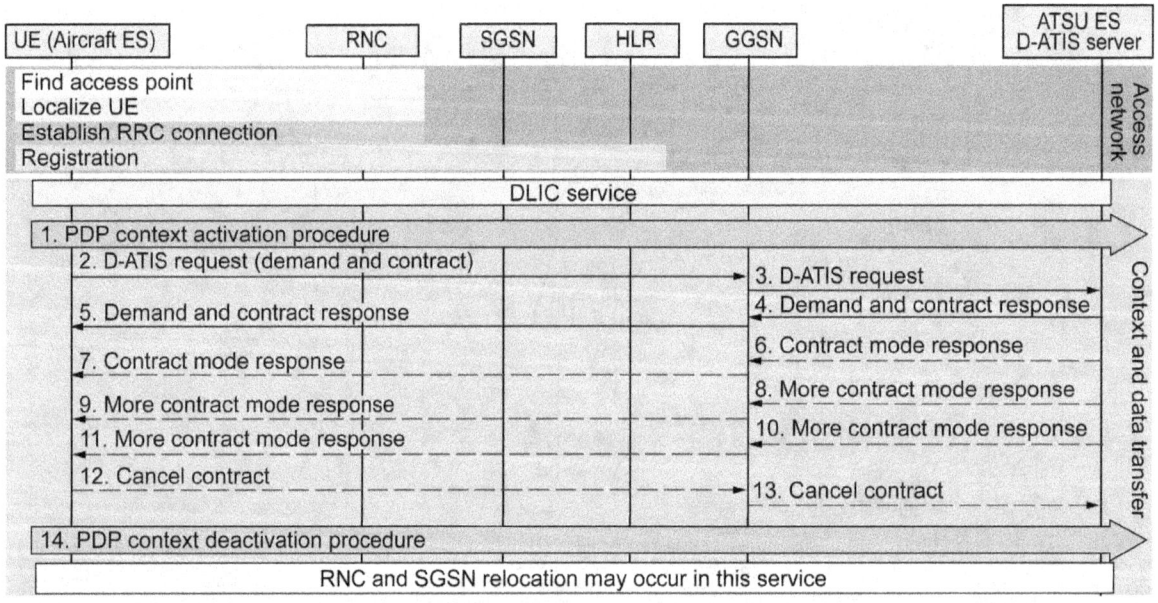

Figure 58.—D-ATIS in WCDMA context.

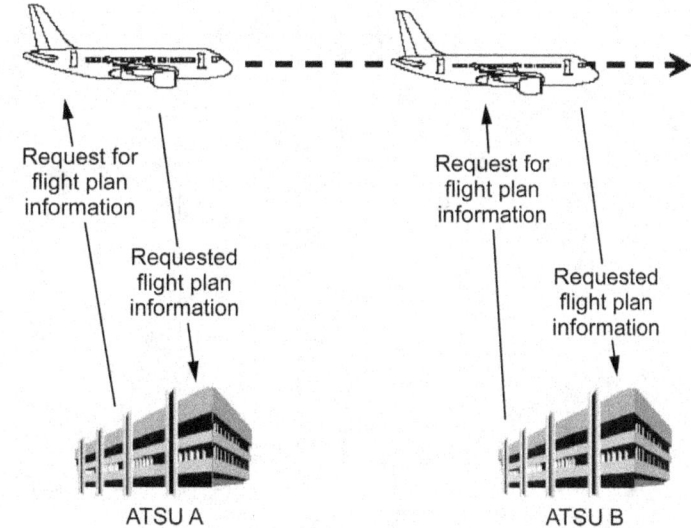

Figure 59.—FLIPCY operational context.

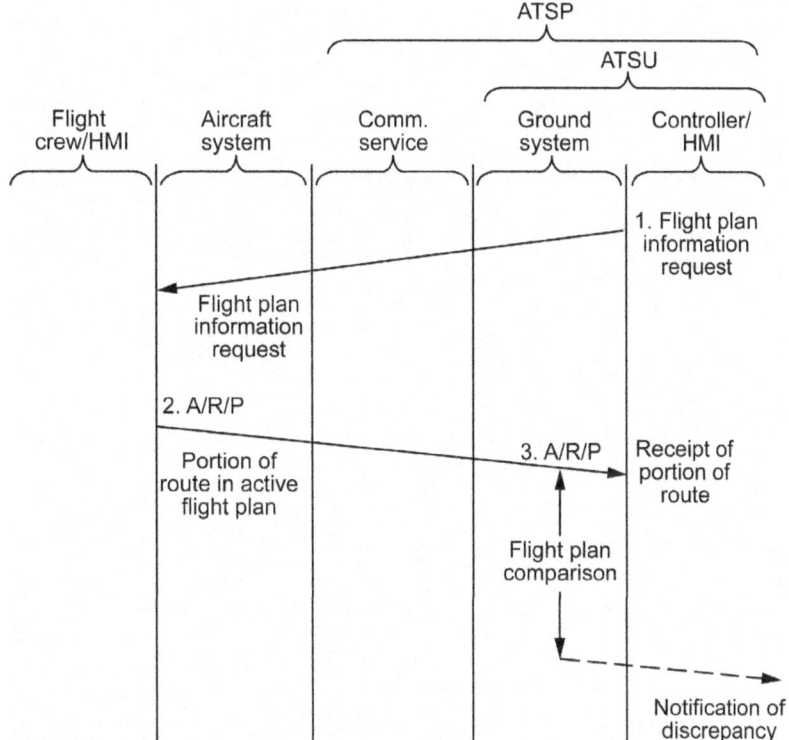

Figure 60.—FLIPCY time sequence.

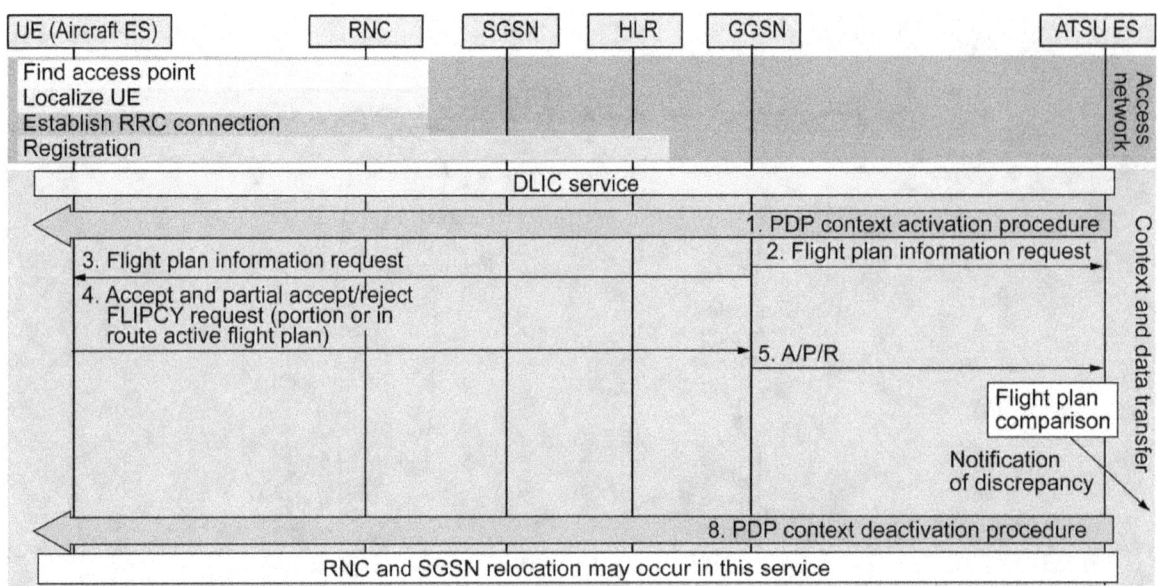

Figure 61.—FLIPCY in WCDMA.

Appendix C—Summary of Previous Indepth Assessments

Many of indepth technical assessments supporting the FCS technology assessment were performed during the FCS Phase II summary. An overview of the work performed and results is provided in the following subsections:

- Section C.1: L-Band Air/Ground (A/G) Channel Characterization
- Section C.2: P34/TIA–902 Performance Assessment
- Section C.3: LDL Performance Assessment
- Section C.4: L-Band Technology Cost Assessment for Ground Infrastructure
- Section C.5: L-Band Interference Analysis
- Section C.6: Satellite Technology Availability Performance
- Section C.7: C-Band Technology (Institute of Electrical and Electronics Engineering (IEEE) 802.16e) Performance

C.1 L-Band Air/Ground (A/G) Channel Characterization

To support the assessment of technology performance in the L-Band Air-Ground (A/G) channel, a literature search revealed that while many channel models exist for the terrestrial channel in close proximity to L-band, there had been no previous activity to develop a channel model that characterizes the L-band A/G channel for radio communications. As most standardization bodies consider it a best practice to test candidate waveform designs against carefully crafted channel models that are representative of the intended user environment, a channel model was developed that could be used for common characterization of communications waveform performance in this A/G channel.

Characterization of the delay spread and the Doppler Power Spectrum is essential for generating a useful model for waveform simulation and evaluation of candidate FRS technologies in L-band. In order to form estimates of the delay spread and associated statistics, a ray-tracing simulation was developed. This simulation models both diffuse and specular reflections from the Earth's surface. The developed simulation used a method of concentric oblate spheroids to model multipath contributions. The desired product was the set of points on the terrain that were intersected by the oblate spheroids. When plotted, each set of intersection points appears as a distorted annulus approximating the cross section of the spheroid when sliced by the Earth's surface. Each set of intersection points is mutually exclusive from any other set because any intersection point can only be accounted for once. Each set of intersection points contributes to multipath for a particular delay. Figure 62 illustrates the method of concentric oblate spheroids used to model multipath contributions.

Implementing the methodology to apply ray tracing to determine specific specular and diffuse multipath components and employing data reduction and analysis techniques, the root mean square (RMS) delay spread was calculated to be 1.4 µs. It is instructive to consider representative technologies at this point since the technology data rate will drive channel model parameter estimation. A rule of thumb frequently applied is that if the mean RMS delay spread is at least one-tenth of the symbol duration, then the channel is frequency-selective. In order to illustrate this, two technologies emerging from the FCS Phase I study were considered: LDL and P34/TIA–902. Table V shows the corresponding data rates and symbol durations for LDL and P34/TIA–902.

TABLE V.—DATA RATES OF LDL AND P34/TIA–902

Waveform	Data rate, kbps R	Symbol duration, µs $T_b = \dfrac{1}{R}$	1/10 of the symbol duration, µs $t_0 = T_b /10$
LDL	62.5	16.0	1.60
P34/TIA–902	4.8[a]	208.3	20.83

[a]P34/TIA–902 is an OFDM system. The tabulated data rate is per carrier and is the symbol rate. Overall P34/TIA–902 data rates range from 76.8 to 691.2 kbps.

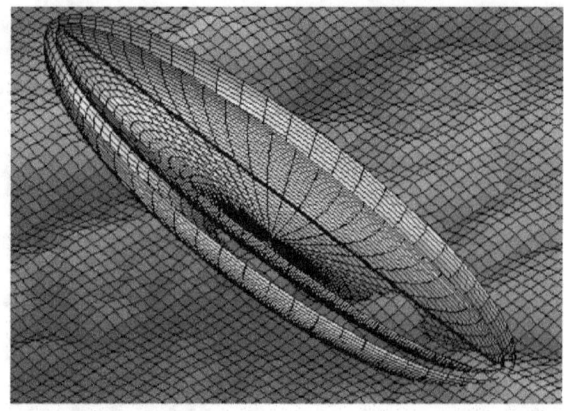

Figure 62.—Two concentric oblate spheroids intersecting the underlying terrain.

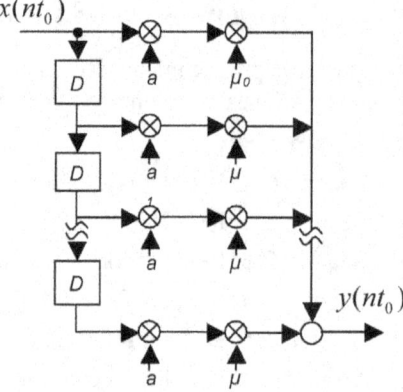

Figure 63.—Block diagram for frequency-selective channel model.

Using our vision, P34/TIA–902 should undergo flat fading and LDL presents a borderline case because the mean RMS delay spread is very close to one-tenth of the symbol duration. It is important to note that frequency-selective channel models differ in structure from flat fading channel models. For this reason it was decided to develop a frequency-nonselective fading model for P34/TIA–902 and a frequency-selective fading model for LDL.

First the channel model for LDL is described. Figure 63 shows the block diagram representation for a deterministic simulation model for a frequency-selective mobile radio channel (Pätzold 270).

The parameters that define the LDL channel model are as follows:

- Number of taps (N)
- Tap spacing (a_0, a_1, ..., a_N)
- Tap weights (D_1, D_2, ..., D_N)
- Tap fading processes (μ_0, μ_1, ..., μ_N)

Table VI defines the LDL channel model parameters that were derived in the Phase II study.

TABLE VI.—LDL CHANNEL MODEL PARAMETERS[a]

Tap no.	Delay, µs	Power, linear	Power, dB	Fading process	Doppler category
1	0.0	1.0000	0	Ricean	Jakes
2	1.6	0.0359	−14.5	Rayleigh	Jakes
3	3.2	0.0451	−13.5	Rayleigh	Jakes
4	4.8	0.0689	−11.6	Rayleigh	Jakes
5	6.4	0.0815	−10.9	Rayleigh	Jakes
6	8.0	0.0594	−12.2	Rayleigh	Jakes
7	9.6	0.0766	−11.2	Rayleigh	Jakes

[a]Note that the assumptions used in the development of the Jakes model are not applicable for the anticipated Ricean A/G channel, but this model has been incorporated to provide a conservative estimate of Doppler effects.

The P34/TIA–902 channel model is much less complex than the LDL channel model because the channel is frequency-nonselective. Figure 64 illustrates the P34/TIA–902 channel model.

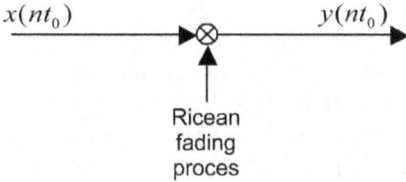

$$x(nt_0) \qquad\qquad y(nt_0)$$

Ricean
fading
proces

Figure 34.—P34/TIA–902 channel model.

The Ricean fading process is derived in the complex baseband by creating two colored Gaussian processes. The Rice method is used to generate the gaussian processes as a summation of sinusoids whose coefficients and frequencies are determined by the Doppler Power Spectrum of the channel. As the process is Ricean, a time-variant mean is summed with the colored gaussian process (loss of signal (LOS) component). The magnitude of the complex gaussian colored processes yields the Ricean process with fade durations and amplitudes determined by the channel.

One of the primary results reported is the simulated RMS delay spread. It should be noted that this delay spread can be modeled as a function of the average distance from the transmitter, with increasing delay spreads reported for increasing distances. Because of this phenomenon, our simulated positions were constrained to be in an area with dimensions that were small compared to the average distance from the transmitter. For these simulations, an RMS delay spread of 1.4 μs was predicted for a certain distance (average distance = 40 miles) from the transmitter in mountainous terrain. A generalized model, using the method cited in Greenstein, has the form

$$\overline{\sigma}_\tau = \overline{\sigma}_{\tau_0} d^\varepsilon A$$

where
    d is the distance in km
    σ_0 is the median value of the RMS delay spread at d = 1 km
    ε is an exponent that lies between 0.5 and 1.0, based on the terrain type
    A is a lognormal variate

To determine the parameters that are appropriate for a generalized L-band A/G model in mountainous terrain, RMS delay spreads were predicted for a reference distance of 1 km as well as for the previously mentioned values at 64.37 km (40 miles). The two predicted values that resulted from the simulation work are

$$\sigma_{RMS}(1 \text{ km}) = 0.1 \text{ μs}$$
$$\sigma_{RMS}(64.37 \text{ km}) = 1.4 \text{ μs}$$

Fitting the Greenstein model to the reference data provides a generalized expression for RMS delay spread, which is found to be

$$\overline{\sigma}_\tau = 0.1 \times d^{0.6337} \quad \text{μs} \quad (A = 6dB)$$

A full description of the L-band A/G channel characterization work is provided in the interim FCS Phase II report, Section E.1.

C.2   P34/TIA–902 Performance Assessment

In addition to L-band channel characterization, L-band technology performance studies specific to individual technologies were also conducted. An indepth analysis of P34/TIA–902 net entry, data transfer, and bit error rate (BER) performance in the L-band channel was performed. The simulation of P34/TIA–902 included evaluation of a ground station and 95 mobile nodes (COCR-defined NAS super sector) employing P34/TIA–902 SAM physical layer properties associated with 50-kHz channelization and QPSK modulation. Simulation model results are shown in figure 65. These figures show the response time of the P34/TIA–902 simulation to the offered load for each of transmitted message. Note that the subnetwork latencies over P34/TIA–902 protocols (subnetwork dependent control protocol (SNDCP), LLC CP, LLC UP, and MAC) meet COCR latency requirements. Specifically, although there are some startup outliers, 95 percent of delay measurements are under 0.7 s. Reference the interim FCS Phase II report, Section E.1.2.2.

In addition to simulation of P34/TIA–902 net entry and data transfer performance, P34/TIA–902 performance in the defined L-band A/G channel was also evaluated. As part of this effort, P34/TIA–902 transmitter and receiver models were generated. Specifically, the P34/TIA–902 SAM physical layer interface was modeled by developing a custom application using C code. The transmitter was implemented as detailed in the specification for the 50-kHz channel using QPSK modulation. The receiver implementation was tested against published results for standardized channel models. Additionally, P34/TIA–902 coding gain (for specified concatenated Hamming codes) was investigated. It was found that a 3×10^{-3} raw BER is approximately equal to 10^{-5} coded BER for P34/TIA–902.

The developed P34/TIA–902 transmitter and receiver models were combined with a model of the expected L-band channel based on analysis work previously described. Specifically, a two tap channel model was simulated where Tap 1 was modeled as Ricean, with a K-factor of 18 dB, unity gain, and Jakes Doppler Spectrum; and Tap 2 was modeled as Rayleigh, with a 4.8 μs delay, –18 dB average energy, and Jakes Doppler Spectrum (conservative estimate). In this model, the mobile velocity was taken to be 0.88 mach. This is the maximum domestic airspeed given in the COCR based on Boeing 777 maximum speed. Additionally, in the model the P34/TIA–902 tuned frequency was taken to be 1024 MHz, with maximum Doppler shift of 1022 Hz.

Initial simulations indicate good performance can be achieved in the aeronautical channel (e.g., flat channel effects), primarily a consequence of the strong line-of-sight component of the received signal (with K factors greater than 4). Figure 66 shows initial performance results and a complete description of the P34/TIA–902 performance assessment including simulation block diagrams, receiver implementation details (channel estimation using interpolation weighting), model validation results, and details on the simulation A/G channel are provided in Section E.1.2.3 of the interim FCS Phase II study report.

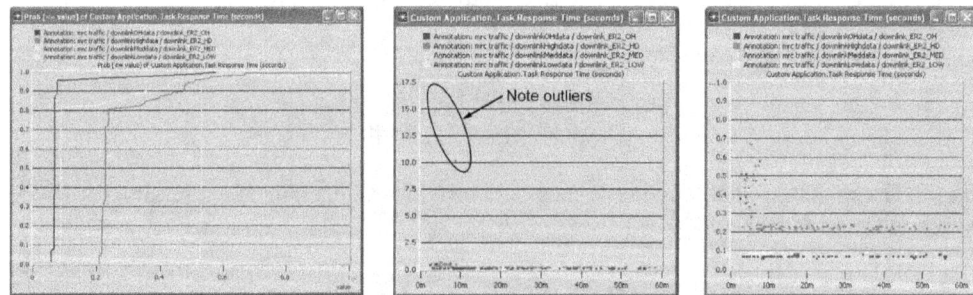

Figure 65.—P34/TIA–902 OPNET modeling results.

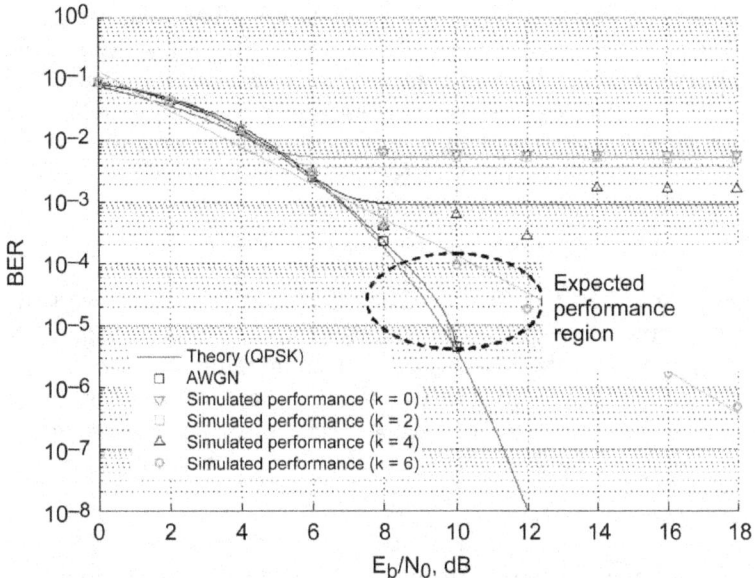

Figure 66.—P34/TIA–902 predicted performance in the L-band aeronautical channel.

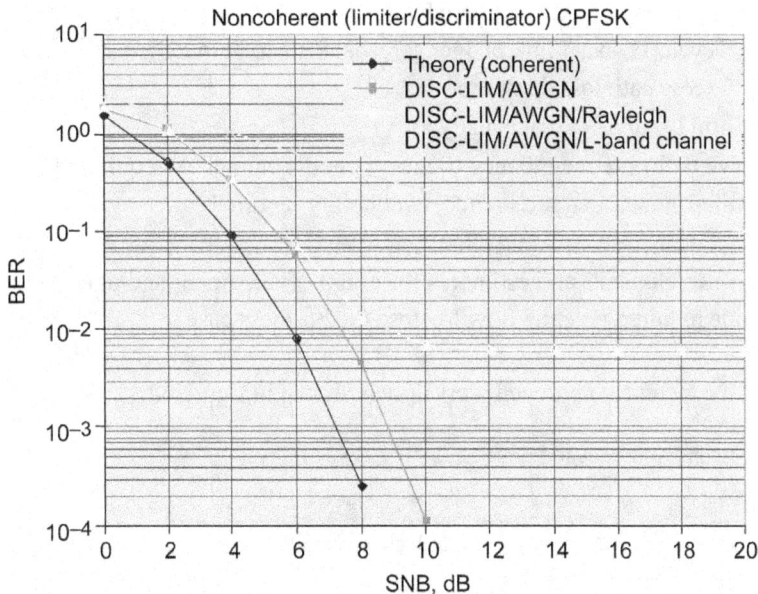

Figure 67.—LDL predicted performance in AWGN and the L-band aeronautical channel.

C.3   LDL Performance Assessment

A second technology investigated for performance in the L-band aeronautical channel was LDL. As with P34/TIA–902, LDL transmitter and receiver models were generated and the receiver model validated against known results. After validation, investigation of LDL coding, Reed-Solomon (72, 62), provides a coding gain of 3 to 4 dB in the expected region of operation. LDL performance was simulated in the L-band aeronautical channel environment. The LDL channel model is a conservative model that introduces an irreducible error floor to system performance (reference Section C.1). The plot shown in figure 67 shows the system performance of LDL in the presence of both additive white gaussian noise (AWGN) and the L-band aeronautical channel model. Based on the results of this simulation, LDL may require channel equalization to mitigate the effects of the A/G aeronautical channel model in L-band.

A complete description of the LDL performance assessment including simulation block diagrams, receiver implementation details, model validation results, and details on the simulation A/G channel are provided in Section E.1.3.2 of the interim FCS Phase II study report.

C.4   L-Band Technology Cost for Ground Infrastructure

L-band technology cost was another focus area of indepth analysis. In this work, the economic feasibility from the perspective of the ground infrastructure provider was evaluated. This analysis was responsive to feedback received on the technology prescreening results (FCS Phase I) that indicated that due to cost constraints, an L-band solution is only considered should VHF spectrum prove insufficient to provide total required data link capability. The L-band business case analysis provided a first order of magnitude estimate of required investment for an L-band aeronautical ground infrastructure. The technical approach for accomplishing this objective included

- Through detailed analysis, develop a notional ground L-band architecture that can meet Future Communication Infrastructure requirements as defined in the COCR document for ATC communications
  - Derive number of radio sites required for total U.S. coverage
    - o Perform L-band link budget analysis
      - □ Develop L-band link budget spreadsheet and derive the parameters to close the link
      - □ Excess path loss derivation
    - o Perform L-band coverage analysis
    - o Derive radio site redundancy to meet system availability requirements
    - o Develop an architecture to meet availability required
- Determine if the business case can close
  - Develop cost elements and estimates for initial development and operations and maintenance
  - Determine required revenue flow to close the business case

An overview of the technical approach workflow is shown in figure 68.

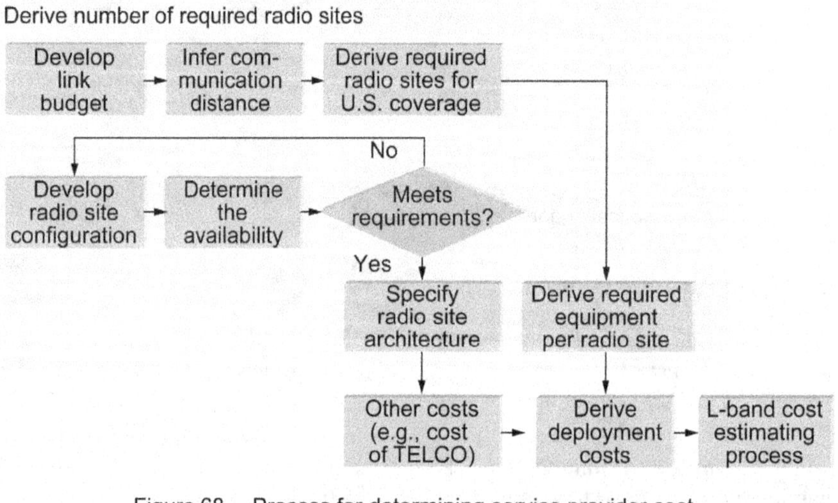

Figure 68.—Process for determining service provider cost.

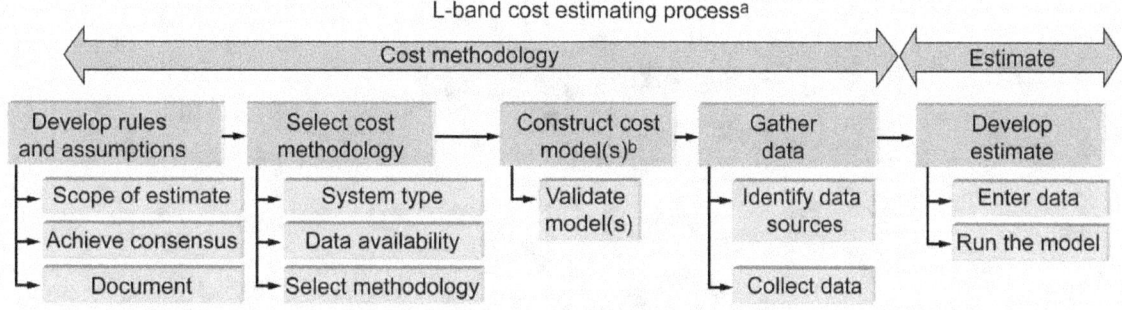

L-band cost estimating process[a]

aBased on NASA Cost Estimating Handbook.
bWilliam G. Sullivan, James A. Bontadelli, Elin M. Wicks: "Engineering Economy," 11th Ed., 2000.

Figure 69.—L-band cost estimating process.

Details of the work performed to develop a link budget, determine radio site redundancy and defined the radio site equipment is provided in the interim FCS Phase II study report (Sections E.1.8.1 and E.1.8.2). The details of the cost estimating approach are shown in figure 69.

Applying the approach above for the L-band cost modeling work, several assumptions were considered including

- L-band provides coverage to a large continental region (e.g., United States or Core Europe)
- Coverage is above FL180
- System Availability of Provision (A_p) meets COCR requirements for COCR Phase II en route services (sans Auto-Execute service)
- Cost elements considered include Research and Development (including system design and engineering); Investment (including facilities and equipment); and Operations and Maintenance (including telecommunications, personnel, and utility costs)

The first order of magnitude cost estimate for implementing an L-band aeronautical ground infrastructure considering life cycle costs and applying the present worth simple payback method (with minimum attractive rate of return = 5 percent) indicates that a positive business case can be achieved (payback period of 4 years). While the first order of magnitude cost estimate yielded positive results, the important aspect of the study to bring forward was the framework of the analysis, which can be considered a generic framework specifying infrastructure costs associated with an L-band system. Additional details specific to the L-band cost assessment are provided in the FCS Phase II report, Section E.1.8.

C.5   L-Band Interference Analysis

Work on this topic was addressed in two ways. First, a modeling effort was performed to assess the interference effects of two candidate technologies on existing L-band aeronautical systems. This work was conducted as part of the FCS Phase II study. Next, as a result of the Phase II findings, an interference measurement campaign was conducted as part of the FCS Phase III study. A summary of the results relating to the analytical assessment are provided here, with details provided in the FCS Phase II report, Section E.1.4. A full description of the objectives, methodology, and results relating to the interference measurement work conducted during FCS Phase III are provided in Section 2.

A candidate spectrum band for the future aeronautical communication radio system is the aeronautical L-band spectrum. This band, 960 to 1215 MHz, has a primary allocation for Aeronautical Radio Navigation Services. There are current several system implementations that occupy the band.

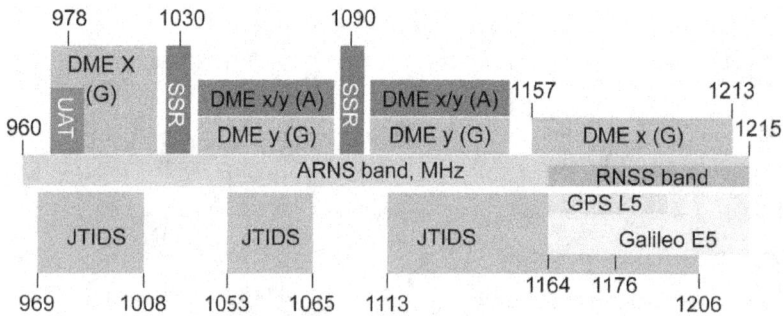

Figure 70.—Current and planned L-band utilization (ref 1).

ICAO systems that use spectrum in this band include the universal access transceiver (UAT); secondary surveillance radars (including Air Traffic Control Radar Beacon System (ATCRBS), Mode A and C, and Mode S); and DME. A majority of the spectrum allocations for these systems are standardized by ICAO. There are, however, some exceptions such as DME allocations defined on a national basis between 962 and 977 MHz in the United States.

Additional systems operating in the aeronautical L-band spectrum include military systems. These include TACAN and joint tactical information distribution system (JTIDS)/multifunctional information distribution system (MIDS) (Link-16). The use of military systems in this band is subject to national coordination between military and civil authorities. Global Navigation Satellite Systems also occupy this band. Specifically, the upper part of the band has been designed for radio-navigation satellite service (RNSS). A visual depiction of the current and planned L-band utilization is shown in figure 70.

As part of the consideration of new future communication system technology implementations in this band, the need to analyze the interference potential of proposed technologies to systems current operating the aeronautical L-band spectrum has been identified. A generic process for interference analysis would have the following elements:

- Describe the source of interference and the interference mechanism
  - Description is usually in the form of power spectrum and time characteristics (e.g., transmit (Tx) power, transmission bandwidth, and duty cycle)
- Quantify the isolation between transmitter output and receiver input
  - This isolation includes the effects of antenna gains, cabling losses, and propagation
- Determine the ratio of desired to undesired (D/U) signal power at the input of the receiver decision process (detector)
- Quantify receiver performance as a function of this D/U ratio, ascribe a required performance, and assess compatibility

The last item noted above is the most difficult element of the process and was the focus of the interference simulation work defined for this study. Specifically, during consensus FAA, NASA and ITT deliberations at the beginning of the Phase II study, two technologies were selected for detailed analysis, LDL and P34/TIA–902. At that time, it was determined that the compatibility of those two proposes systems with existing ICAO standardized civil aviation systems would be included in the detailed analysis. Thus, the objective of the interference analysis task was to determine the compatibility of P34/TIA–902 and LDL with standardized civil aviation systems. The approach for the interference analysis included (for each system being analyzed)

- Selection of an appropriate measure of interference degradation
- Collection of information about the system (known susceptibilities and system technical parameters)

- Development of a physical layer system model and validation with known results
- Introduction of the interference source and prediction of victim performance

In an effort to prioritize analysis resources, a list of individual candidate interference analyses was defined. This list is provided in table VI.

TABLE VI.—CANDIDATE INTERFERENCE ANALYSES

Interference source	Victim receiver	Interference mechanisms	Source characterization	Has vulnerability been characterized?
FRS 960 to 1024 MHz	GNSS	Broadband noise	Noise (WB)	Yes
		Spurious emissions	NB or CW	Yes
	1176.45 MHz	Desensitization		
960 to 977 MHz Preferred				
	Mode S	Broadband noise	Noise (WB)	Yes
		Spurious emissions	NB or CW	Unknown
	1030 MHz	Desensitization		
	1090 MHz			
	UAT	Adjacent signal	FRS dependent	No
	978 MHz	Broadband noise	Noise (WB)	Yes
		Spurious emissions	NB or CW	Yes
	DME	Co-channel	FRS dependent	No
	962 to 1019 MHz	Adjacent signal	FRS dependent	No
		Broadband noise	Noise (WB)	Yes
		Spurious emissions	NB or CW	yes

In the table above, it can be noted that some of the vulnerabilities have previously been characterized. Therefore, the focus of this study was on the vulnerabilities in red, that is, those vulnerabilities that have not previously been addressed.

C.5.1 DME Interference Assessment

The DME system is an ICAO-standardized navigational aid used to determine the aircraft location. It consists of an interrogator located onboard the aircraft and a transponder located at a ground station. At regularly spaced intervals, the interrogator transmits a coded pulse to the transponder. Reception of this pulse triggers a coded reply from the interrogator at a different frequency. The DME system uses the principle of elapses time measurement between these two messages as the basis for determining the distance between the aircraft and the ground station, also called the slant range distance. DME frequencies are spaced in 1 MHz increments throughout the 962 to 1213 MHz band, providing potential for interference to and from FRS in L-band. A list of known susceptibilities and previous DME susceptibility test were reviewed.

With the knowledge gained from review of existing DME tests, a first step in the FCS/DME interference analysis was the development of a DME receiver model. To perform this work, published data on DME interference from global positioning system (GPS) signals was obtained. The data indicated that interference from P(Y) and from C/A signals does not differ much, even though the P(Y) signal has 10 times larger bandwidth. Thus a hypothesis was developed, which assumes that pulse detection in DME equipment is performed over a short window, on the order of one P(Y) chip length. A receiver window length, which would yield a match with the published data, was then computed.

This hypothesis and associated DME architecture assumptions were applied and a mathematical model, which describes this architecture, was built. The model was run for different values of parameters to determine sets of parameters that matched published results. The implemented model was then tested using a UAT interfering signal to test results. The developed model and associated UAT test results are shown in figure 71.

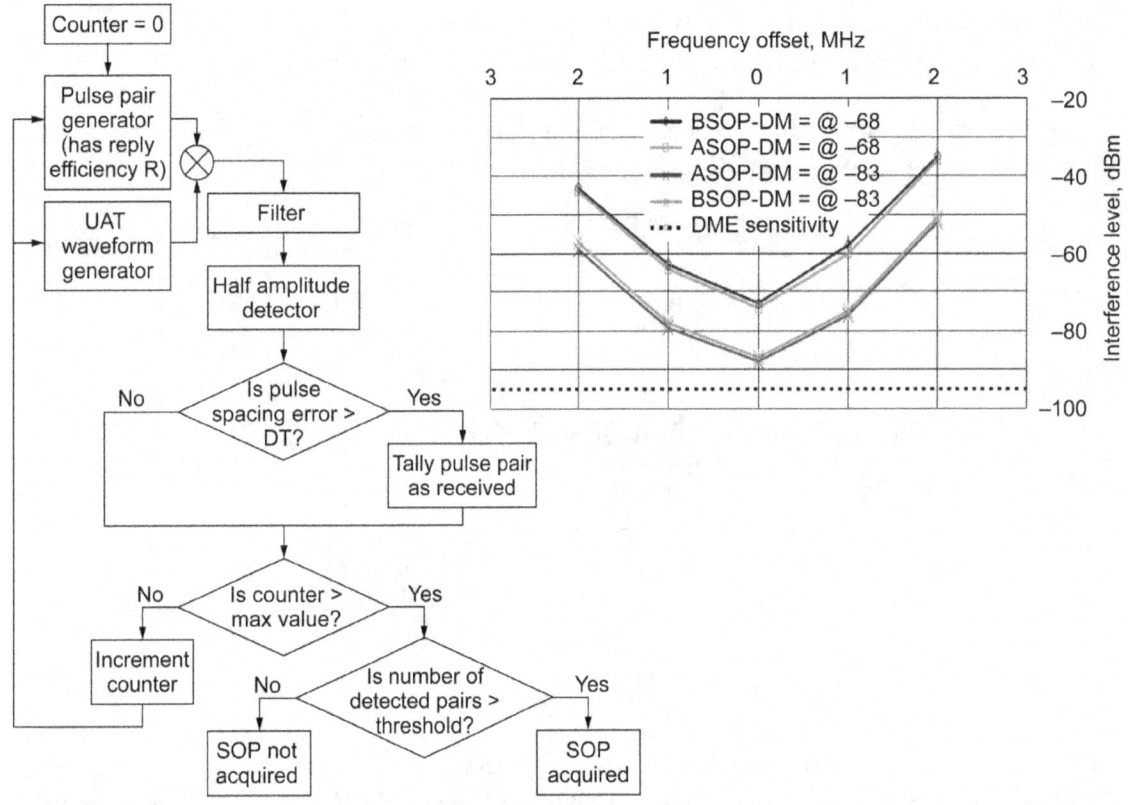

Figure 71.—Implemented DME model and UAT interference results.

For the model above, DME pulses were modeled as gaussian. UAT was modeled as a frequency-shift keying, constant amplitude signal. Here, the DME pulses and interference were superimposed in the time domain. The resulting signal was filtered using a filter with gaussian response function; the width of the filter response is computed to match a measured decrease of interference effect as a frequency offset of 1 MHz as compared to no offset.

Results captured were compared with published data and despite the seemingly good correlation of results of the developed model and measurements, several problems with the developed model were noted during validation testing. Specifically,

- The measured results are extremely flat over the reply efficiency range of the test.
  - Indicative of an automatic gain control circuit (perhaps) or some second order effect that is not immediately obvious
- To create a range of "Acquire Locks" for various reply efficiencies, the interference power for our model had to be varied over a 10 to 12 dB range.
  - This was deemed to be sufficiently far from measured results as to be a unreliable indicator (for use in predicting interference from FRS sources).
- Several requests for information and assistance were made by NASA, but the information that was needed (detailed algorithm descriptions from radio manufacturers) was not made available.

As a result of the observations above, a decision was made to not further use the developed model. Rather, measurements were recommended to more substantively characterize the DME to communication waveforms in the final phase of the FCS technology assessment.

C.5.2 UAT Interference Modeling

UAT is a wideband data link that enhances pilot situation awareness and increases safety by allowing general aviation pilots to process navigational signals from the GPS, receive traffic information, broadcast their position, and perform other functions. It is a technology that is standardized through ICAO for Automatic Dependent Surveillance, Broadcast (ADS–B); Traffic Information Services, Broadcast (TIS–B); and Flight Information Services, Broadcast (FIS–B). UAT operates at 978 MHz, providing potential for interference to and from a FRS in L-band.

UAT has several known susceptibilities. These include

- DME signal interference (basic and high-performance receivers)
  - 99 percent successful message reception of long messages in presence of DME pulse pairs at a nominal rate of 3600 ppps at either 12 or 30 μs spacing at a level of –30 dBm for any 1 MHz channel frequency between 980 and 1215 MHz (desired signal >= –90 dBm)
- DME signal interference (basic receivers only)
  - 90 percent successful message reception of long messages in presence of DME pulse pairs at a nominal rate of 3600 ppps at either 12 or 30 μs spacing at a level of –56 dBm for any 1 MHz channel frequency between 979 MHz (desired signal >= –87 dBm)
- DME signal interference (high-performance receivers only)
  - 90 percent successful message reception of long messages in presence of DME pulse pairs at a nominal rate of 3600 ppps at either 12 or 30 μs spacing at a level of –43 dBm for any 1 MHz channel frequency between 979 MHz (desired signal >= –87 dBm)

For this study, the objective was to characterize the impact of LDL and P34/TIA–902 interference on UAT performance. To perform the analysis, several assumptions were employed. For UAT, the basic ADS–B message code RS(30,18) has been modeled. And, the analysis did not include long ADS–B message codes RS(48,34) or Ground Uplink Message Codes RS(92,72). For LDL, the transmitter model used a data rate of 62.5 kbps. The analysis did not consider other possible LDL data rates. And finally, for P34/TIA–902, the 50-kHz channelization configuration of P34/TIA–902 was modeled. The analysis did not consider the 100- or 150-kHz configurations.

Details of the UAT/LDL/P34/TIA–902 transmitter block diagrams; analysis parameters and SPW tool transmitter implementations are described in Section E.1.4.2 of the FCS Phase II report. The end-to-end simulation model is shown in figure 72.

The model was validated using the AWGN environment and good correlation with published results achieved. A summary of simulation results, which include a collection of BER curves for varying degrees of LDL/P34/TIA–902 interference into the UAT signal, are shown in figure 73.

From the curves above, it would appear that a carrier to interference (C/I) ratio between 12 and 15 dB is required for minimum degradation to the UAT receiver. LDL has slightly better performance than P34/TIA–902 in terms of not interfering with UAT receivers.

C.5.3 Mode S Interference Modeling

Mode Select (Mode S) is a system developed to phase out the ATCRBS by providing enhanced surveillance information for use by air traffic control automation. Mode S provides more accurate position information and minimizes interference by discreet interrogation of each aircraft. Each aircraft has its own unique Mode S address, providing a mechanism by which an aircraft can be selected and/or interrogated such that no other aircraft reply. Mode S also provides a digital data link to exchange information between aircraft and various ATC functions and weather databases. The system operates at 1030 and 1090 MHz providing a potential for interference to and from a FRS in L-band.

The developed Mode S transmitter simulation model exactly met the rise-time, decay-time, and power spectral density (PSD) mask requirements given in the Mode S MOPS. The developed simulation modeled the Mode S preamble detection circuit, making a hard decision on every 0.5 μs symbol.

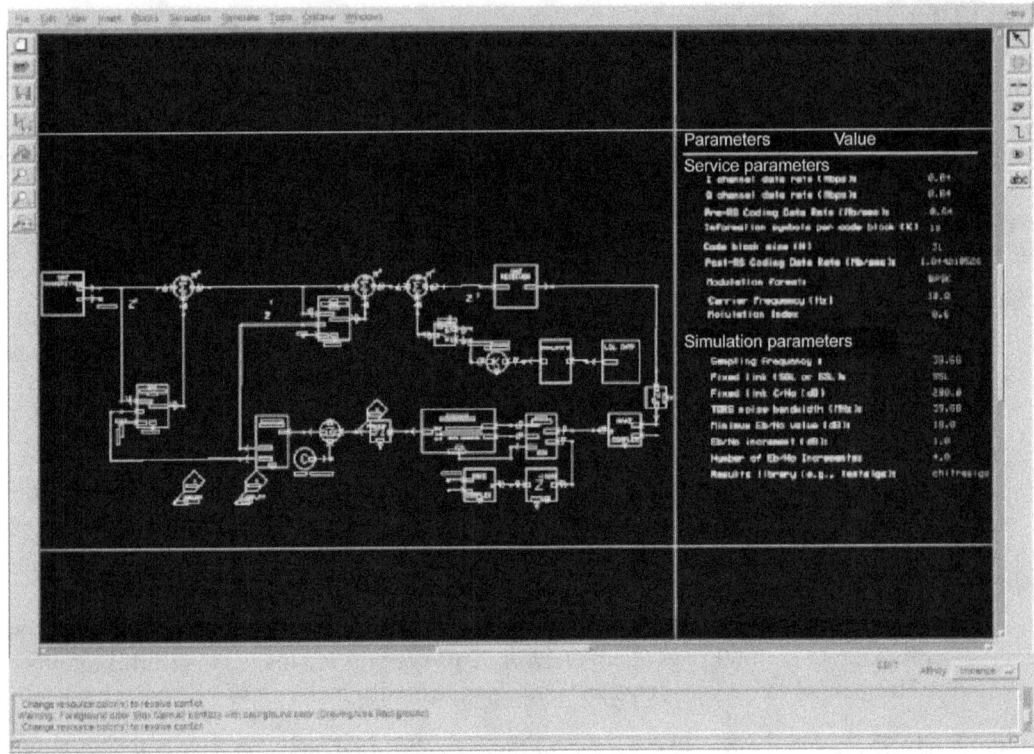

Figure 72.—UAT end-to-end simulation model.

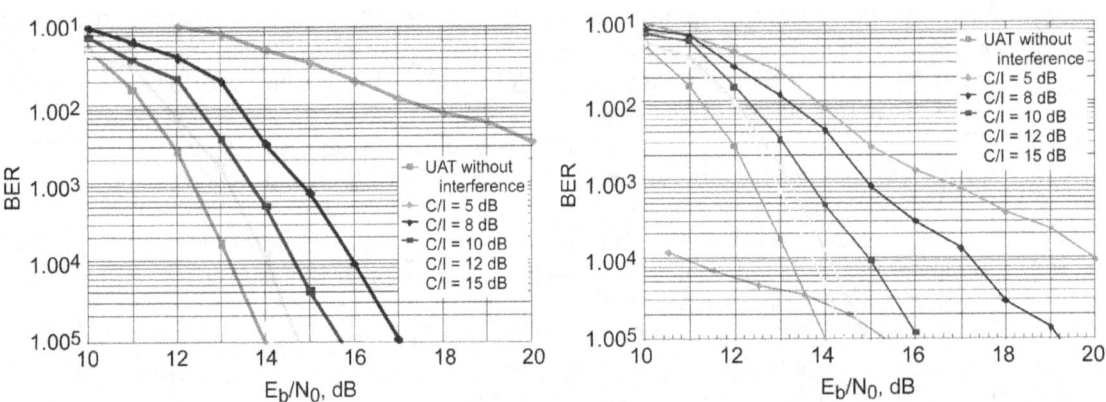

Figure 73.—UAT interference assessment results.

Selectable sensitivity is also included in the model. Using the developed Mode S transmitter and preamble detection models, an end-to-end simulation was created. This end-to-end model included integrated LDL and P34/TIA–902 interferer models. Details on the simulation block diagrams and SPW transmitter implementations are described in Section E.1.4.3 of the FCS Phase II report.

For both LDL and P34/TIA–902 interferences, Mode S probability of correct preamble detection and probability of false preamble detection were measured for several interference levels and several assumptions of required correlation to achieve preamble detection. To compare the interfering effects of P34/TIA–902 and LDL, a probability of correct preamble detection based on varying C/I values (for 94 percent correlation and 100 percent correlation for declaring detection) for both P34/TIA–902 and LDL interferers are shown in figure 74.

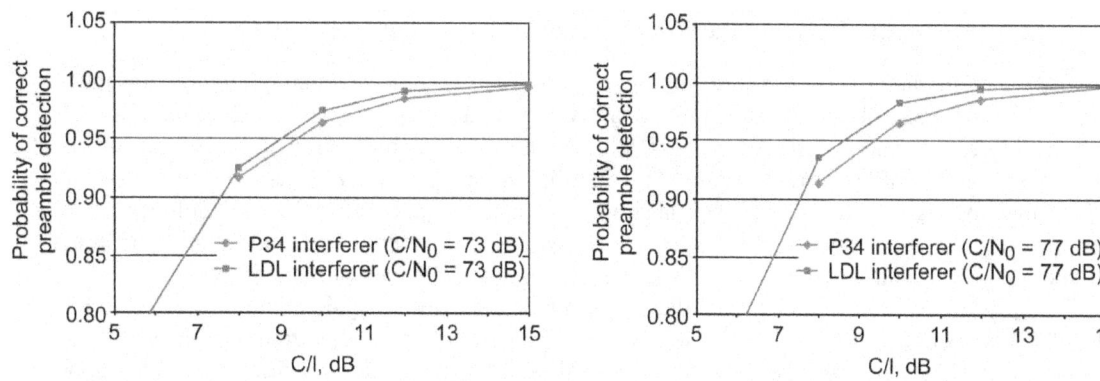

Figure 74.—Comparing effects of P34/TIA–902 and LDL interference on mode S—preamble detection. Note: Conversion from C/N_0 to C/N within necessary bandwidth can be done as follows: C/N within necessary bandwidth = $C/N_0 + 10\log_{10}(2) - 10\log_{10}(4{,}000{,}000)$.

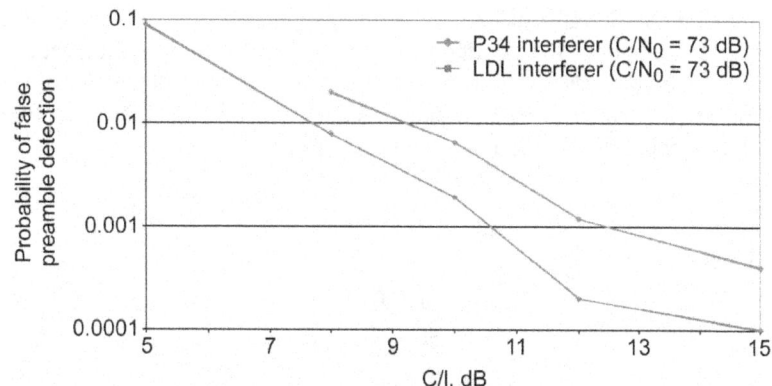

Figure 75.—Comparing effects of P34/TIA–902 and LDL interference on mode S–false preamble detection.

A similar comparison of interfering effects for Mode S probability of false preamble detection is shown in figure 75.

The modeling results would seem to indicate that a C/I ratio of 15 dB or better is required to not substantially degrade the Mode-S preamble detection performance. The behavior of "false preamble detection" would appear to be somewhat worse than the behavior of "missed preamble detection." As in the UAT case, the performance of LDL is better than that of P34/TIA–902; that is, P34/TIA–902 acts as more of an interference source than LDL to both Mode S and UAT receivers. It should be noted that all simulations were made "on-tune;" actual deployment scenarios should be far off-tune, especially for the Mode S case (proposed band for the FRS is 960 to 1024 MHz, and the Mode S Extended Squitter equipment is at 1090 MHz). Additionally, measurements should be made that further characterize Mode S behavior as there are other metrics to investigate besides preamble detection. Finally, the preamble detection modeled is hardly sophisticated, and better performance from actual equipment is predicted.

C.6   Satellite Technology Availability Performance

For the satellite and over horizon technology family, two technology inventory candidates have emerged from the technology screening: Inmarsat Swift Broadband (SBB) and Custom Satellite Solutions. For satellite aeronautical communication solutions, availability typically arises as an important issue to address. In order to provide required availability, a highly redundant custom satellite system architecture is needed. As this issue is similar for both Inmarsat and Custom Satellite Solutions, it was considered instructive to estimate the availability of two existing, operational satellite systems, Inmarsat SBB and Iridium, which provide services in protected aeronautical spectrum (AMS(R)S).

The approach used for SATCOM availability modeling was the analysis model described in RTCA DO–270. This document defines an availability fault tree to permit characterization and evaluation of multiple availability elements. The fault tree is organized into two major categories, system component failures and fault-free rare events. This model, shown in figure 76, was useful for comparing architectures and was applied in this study.

Details of the evaluation of each component failure are documented in the FCS Phase II report, Section E.2. A summary of availability modeling results is shown in figure 77. For SATCOM systems, limiting factors for availability include satellite equipment failures and RF link effects (Inmarsat and Iridium), capacity overload (Iridium), and interference (Iridium). For the VHF terrestrial reference architecture, the limiting factors for availability include RF link events and capacity overload. Overall, the detailed evaluation of satellite communication systems (with a focus on provision of required availability) indicated that both Inmarsat SBB and Iridium would not meet availability requirements. Also, a custom satellite solution designed to meet COCR availability requirements would, in fact, require a highly redundant and costly architecture. Although availability concerns may limit the use of satellites as cost-effective solutions for continental airspace domains, this does not preclude their effective role in providing communication capability in remote and oceanic airspace.

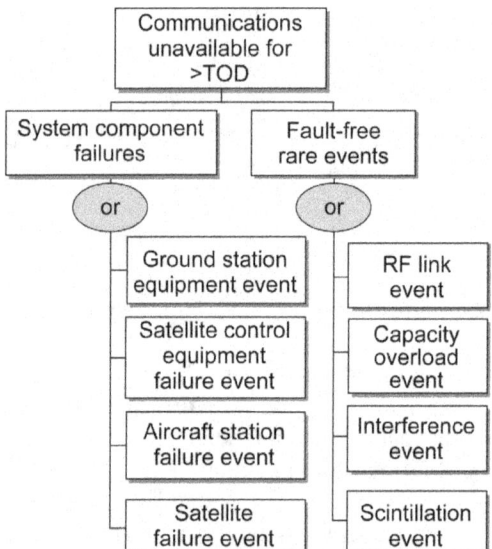

Figure 76.—SATCOM availability modeling
approach—fault tree.

	System component failures				Fault-free rare events			
	Ground station	Control station	Aircraft station	Satellite	RF link	Capacity overload	Interference	Scintillation
Inmarsat	~1	~1	~1	0.9999	0.95	~1	~1	~1
Iridium	0.99997	~1	~1	0.99	0.995	~1	0.996	~1
VHF Terrestrial	0.99999	N/A	~1	N/A	0.999	~2	~1	N/A

Notes:

1. NASA Technology Investigations for the Future Communications Study. Iridium Capacity Overload availability of AES to SATCOM traffic is essentially one (1) (for both ATS only and ATS and AOC). No steady state can be achieved for SATCOM to AES traffic.

2. Terrestrial Capacity Overload availability is for VHF-band reference architecture business case; for L-Band Terrestrial Capacity Overload availability would be essentially one (1).

Figure 77.—Summary of availability modeling results.

C.7   C-band Technology (IEEE 802.16e) Performance

C-band modeling activities were conducted to investigate the utility of an industry standard system in the airport surface environment. The system that was chosen for analysis was the IEEE 802.16e metropolitan area network (MAN) interface standard. The IEEE 802.16e standard (referred to as simply the 802.16e standard, or 802.16e henceforth) was chosen as it scored well during the initial phase (technology prescreening) of the FCS technology investigations.

As the 802.16e standard supports a range of physical layers, prior to the modeling process, a specific physical layer needed to be selected. Of the possible candidates, better mobility performance is expected from OFDMA than OFDM, and the leading commercial 802.16 forum (the WiMAX Forum) has defined "mobile" WiMAX profiles, which are all expected to adopt the OFDMA physical layer. In this study, however, the OFDM physical layer was selected for analysis, as it seems that if good performance can be predicted for OFDM then by inference the OFDMA physical layer would also work well. Further, there are commercially available chipsets for the 802.16 OFDM physical layer currently available. Since a logical next step to this research would be prototype implementations and trials in the band, and noting that OFDM (due to the aforementioned chipset) is more amenable to prototype equipment development, this seemed to be a reasonable decision.

Implementing the methodology defined above, 802.16e transmitter and receiver functions were modeled in the MATLAB Simulink environment. The next step in the C-band modeling work was to validate the developed model, as depicted in figure 78. Specifically, the simulation was executed in an AWGN environment and corresponding results compared to published results. Good correlation was achieved. Details related to the developed models and validation results can be found in the FCS Phase II report, Section E.3.2.1.

Using a channel model adapted from a detailed model developed by Ohio University (described in detail in Section E.3.3 of the FCS Phase II report), the performance of 802.16e in the aeronautical airport environment was simulated as shown in figure 79. Here performance was found to be quite good for most of the movement area (incorporating equalization techniques). While this technology has good potential applicability for this domain, additional analysis to look at features to enhance performance (e.g., hybrid automatic repeat request (HARQ), fast feedback channel and diversity subcarrier permutations) is warranted.

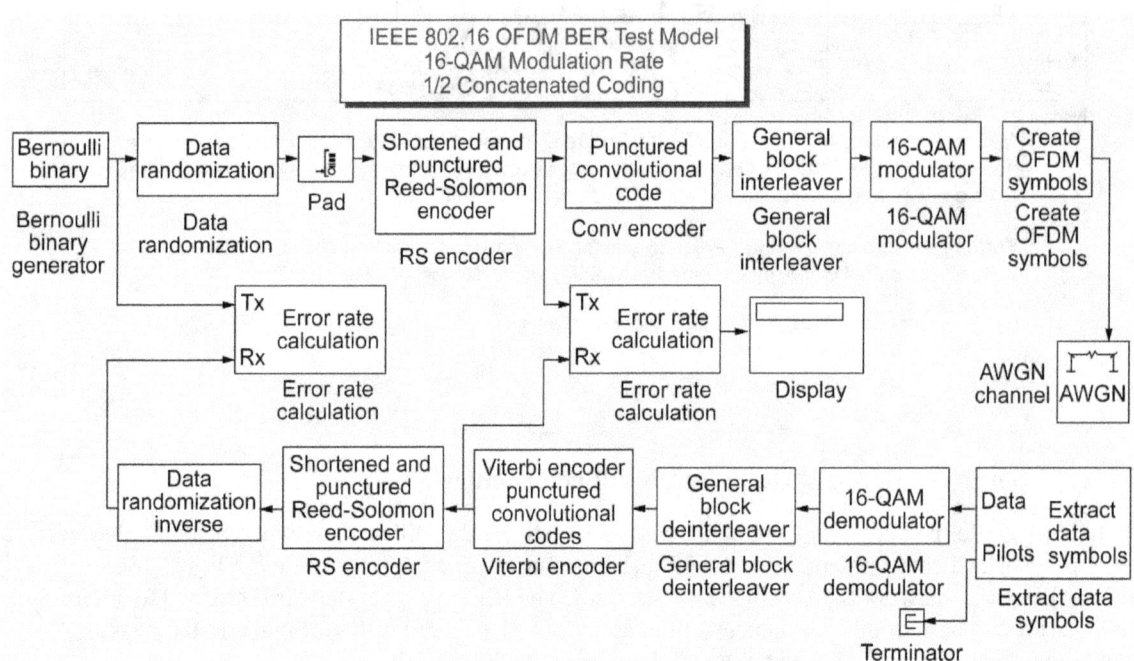

Figure 78.—802.16e end-to-end model.

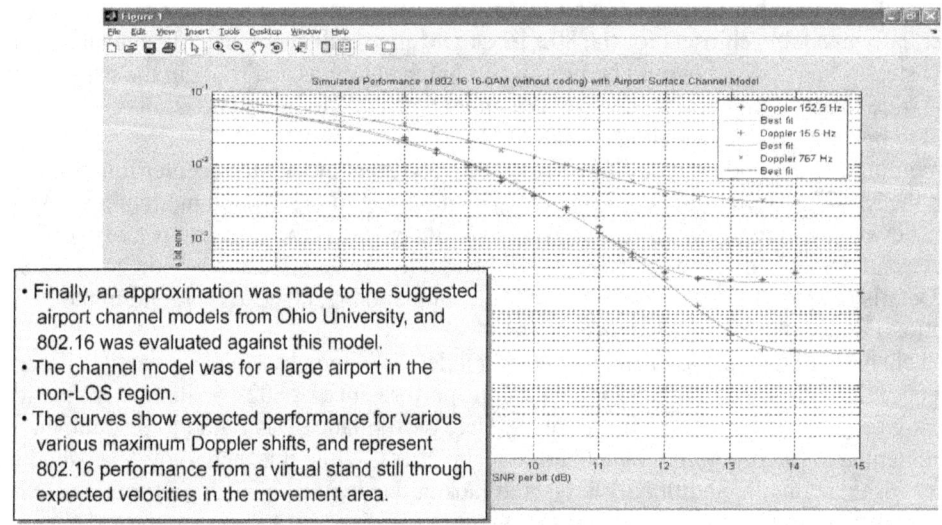

Figure 79.—802.16e simulation results for the aeronautical C-band surface channel model.

Appendix D—List of Acronyms and Abbreviations

3GPP	Third Generation Partnership Project
ACL	air traffic control clearances
ACM	air traffic control clearance message
ACP	Aeronautical Communications Panel
ADS–B	Automatic Dependent Surveillance Broadcast
AMC	ATC microphone check
AMCP	Aeronautical Mobile Communication Panel
AOC	aeronautical operational control
APCO	Association of Public-Safety Communications Officers
ASOP	acquire stable operating point
ATC	air traffic control
ATCRBS	Air Traffic Control Radar Beacon System
ATIS	Automatic Terminal Information Service
ATM	air traffic management
ATMAC	Air Traffic Management Advisory Committee
ATS	air traffic services
ATSP	air traffic service provider
ATSU	Air Traffic Services Unit
AuC	authentication center
AWGN	additive white gaussian noise
BER	bit error rate
BSOP	break stable operating point
CDMA	code division multiple access
CHC	channel coding
C/I	carrier to interference
CN	Core Network
CNS	communication, navigation, surveillance
COCR	Communications Operating Concept and Requirements
CPDLC	controller pilot data link communications
CS	circuit switched
D-ATIS	digital-automatic terminal information service
DCL	departure clearance
DLIC	data link initiation capability
DME	distance measuring equipment
DSC	downstream clearance
D/U	desired to undesired
DUT	device under test
EIR	equipment identity register
FAA	Federal Aviation Administration
FCS	Future Communications Study
FLIPCY	flight plan consistency
FNE	fixed network equipment
FRS	future radio system
GGSN	gateway GPRS support node
GPRS	general packet radio service
GPS	global positioning system
GSM	Global System for Mobile communication
HARQ	hybrid automatic repeat request
HF	high frequency

HLR	home location register
ICAO	International Civil Aviation Organization
IEEE	Institute of Electrical and Electronics Engineering
IOTA	isotropic orthogonal transform algorithm
IP	internet protocol
IPv4	IP version 4
IPv6	IP version 6
JTIDS	joint tactical information distribution system
LDL	L-band digital link
LOS	loss of signal
MAC	Media Access Control
MAN	metropolitan area network
MHz	megahertz
MIDS	multifunctional information distribution system
MSC	mobile switching center
NASTD	National Association of State Telecommunications Directors
NOTAM	notice to airmen
OFDM	orthogonal frequency division multiplexing
PDP	packet data protocol
PDN	packet data network
PHY	physical
PRF	pulse repetition frequency
PS	packet switched
PSD	power spectral density
QAM	quadrature amplitude modulation
QPSK	quadrature phase shift keying
RA	routing area
RLA	Radio Link Adaptation
RMS	root mean square
RNC	Radio Network Controller
RNSS	radio-navigation satellite service
RRC	Radio Resource Control
RTCA	Radio Technical Commission for Aeronautics
RTP	real time protocol
SAM	Scalable Adaptive Modulation
SBB	Swift Broadband
SGSN	serving GPRS support node
SNDCP	sub network dependent convergence protocol
TCP	transmission control protocol
TDMA	time division multiple access
TELCO	Telephone Company
TIA	Telecommunications Industry Association
UAT	universal access transceiver
UDP	user datagram protocol
UE	user equipment
UMTS	Universal Mobile Telecommunications System
URA	Universal Mobile Telecommunications System Terrestrial Radio Access Network Registration Area
USPTO	United States Patent and Trademark Office
UTRAN	UMTS Terrestrial Radio Access Network
VHF	very high frequency

VLR   visitor location register
WCDMA   wideband code division multiple access

References

1. Muraca, Peter: ATMAC Recommendations, January 2005 (documented in Data Communications Program). FAA, 2007.
2. Aeronautical Communications Panel (ACP) Working Group of the Whole, First Meeting. Montreal, Report of the First Meeting, Agenda Item 5: Review of the Progress on the Development of New Communication Systems, ACP–WGW01/AI–5, June 2005.
3. Gilbert, Tricia, et al.: Identification of Technologies for Provision of Future Aeronautical Communications. NASA/CR—2006-214451, 2006.
4. Vega, Elena: Framework for Spectrum Compatibility Analysis in L-Band for FCI Technology Candidates, AENA.
5. Technology Assessment for the Future Communications System. ITT Industries/NASA, May 2005.
6. EMC Analysis of JTIDS in the 960–1215 MHz Band. Mar. 1978.
7. MIL–STD–449D: Measurement of Radio Frequency Spectrum Characteristics. Department of Defense, Feb. 1973.
8. Scalable Adaptive Modulation (SAM) Physical Layer Specification—Public Safety Wideband Data Standards Project—Digital Radio Technical Standards. TIA–902.BAAB–A, TIA, Sept. 2003.
9. L-Band 3G Ground-Air Communication System Interference Study. Roke Manor, Dec. 2006.
10. 3GPP Specifications. http://www.3gpp.org/specs/numbering.htm Accessed Dec. 14, 2007.
11. RTCA/DO–290—Safety and Performance Requirements Standard for Air Traffic Data Link Services in Continental Airspace. RTCA, Inc., Apr. 2004.